Recent Remote Sensing Sensor Applications - Satellites and Unmanned Aerial Vehicles (UAVs)

Edited by Maged Marghany

Published in London, United Kingdom

IntechOpen

Supporting open minds since 2005

Recent Remote Sensing Sensor Applications - Satellites and Unmanned Aerial Vehicles (UAVs)
http://dx.doi.org/10.5772/intechopen.95162
Edited by Maged Marghany

Contributors
Huichun Ye, Wenjiang Huang, Shanyu Huang, Chaojia Nie, Jiawei Guo, Bei Cui, Fei He, Zhonghua Yao, Yong Wei, Ronak Jain, Jiali Shang, Jiangui Liu, Zhongxin Chen, Heather McNairn, Andrew Davidson, Deniz Kumlu, Nur Huseyin Kaplan, Isin Erer, Atri Sanyal, Amitabha Sinha, Hamdan Omar, Thirupathi Rao Narayanamoorthy, Norsheilla Mohd Johan Chuah, Nur Atikah Abu Bakar, Muhamad Afizzul Misman, Yuri V. Kim, Maged Marghany

Notice
Statements and opinions expressed in the chapters are these of the individual contributors and not necessarily those of the editors or publisher. No responsibility is accepted for the accuracy of information contained in the published chapters. The publisher assumes no responsibility for any damage or injury to persons or property arising out of the use of any materials, instructions, methods or ideas contained in the book.

First published in London, United Kingdom, 2022 by IntechOpen
IntechOpen is the global imprint of INTECHOPEN LIMITED, registered in England and Wales, registration number: 11086078, 5 Princes Gate Court, London, SW7 2QJ, United Kingdom
Printed in Croatia

British Library Cataloguing-in-Publication Data
A catalogue record for this book is available from the British Library

Additional hard and PDF copies can be obtained from orders@intechopen.com

Recent Remote Sensing Sensor Applications - Satellites and Unmanned Aerial Vehicles (UAVs)
Edited by Maged Marghany
p. cm.
Print ISBN 978-1-83969-544-5
Online ISBN 978-1-83969-545-2
eBook (PDF) ISBN 978-1-83969-546-9

We are IntechOpen,
the world's leading publisher of
Open Access books
Built by scientists, for scientists

6,000+
Open access books available

148,000+
International authors and editors

185M+
Downloads

Our authors are among the

156
Countries delivered to

Top 1%
most cited scientists

12.2%
Contributors from top 500 universities

Interested in publishing with us?
Contact book.department@intechopen.com

Numbers displayed above are based on latest data collected.
For more information visit www.intechopen.com

Meet the editor

Prof. Dr. Maged Marghany obtained a BSc in Physical Ocean-ography from the University of Alexandria, Egypt, an MSc in Physical Oceanography from the University Pertanian Malaysia, a Ph.D. in Environmental Remote Sensing from the Universiti Putra Malaysia, and a post-doctorate degree in Radar Remote Sensing from the International Institute for Aerospace Survey and Earth Sciences, The Netherlands. The prestigious Universi-dade Estadual de Feira de Santana, Universidade Federal da Bahia, and Universidade Federal de Pernambuco, Brazil ranked him as the first global scientist in the field of oil spill detection and mapping during the last fifty years. Dr. Marghany is currently a director of Global Geoinformation Sdn.Bhd. His research specializes in microwave remote sensing and remote sensing for mineralogy detection and mapping. Previously, he worked as a deputy director in research and development at the Institute of Geospatial Science and Technology and the Department of Remote Sensing, both at Universiti Teknologi Malaysia. He has published more than 250 papers in international conferences and journals and eight books. He is active in the international geoinformatics journal and the International Society for Photogrammetry and Remote Sensing (ISPRS). Dr. Marghany was listed among the world's top 2% of scientists by Stanford University, USA in 2020 and 2021.

Contents

Preface   XI

Chapter 1   1
Introductory Chapter: Automatic Detection of Ice Covers
in Airborne Radar Data Using Genetic Algorithm
by Maged Marghany

Chapter 2   11
Recent Advancement of Synthetic Aperture Radar (SAR)
Systems and Their Applications to Crop Growth Monitoring
by Jiali Shang, Jiangui Liu, Zhongxin Chen, Heather McNairn
and Andrew Davidson

Chapter 3   35
Utilization of Remote Sensing Technology for Carbon Offset
Identification in Malaysian Forests
by Hamdan Omar, Thirupathi Rao Narayanamoorthy,
Norsheilla Mohd Johan Chuah, Nur Atikah Abu Bakar
and Muhamad Afizzul Misman

Chapter 4   55
Optical Remote Sensing of Planetary Space Environment
by Fei He, Zhonghua Yao and Yong Wei

Chapter 5   79
Image Enhancement Methods for Remote Sensing: A Survey
by Nur Huseyin Kaplan, Isin Erer and Deniz Kumlu

Chapter 6   101
Feature-Oriented Principal Component Selection (FPCS)
for Delineation of the Geological Units Using the Integration
of SWIR and TIR ASTER Data
by Ronak Jain

Chapter 7   129
Trans_Proc: A Processor to Implement the Linear Transformations
on the Image and Signal Processing and Its Future Scope
by Atri Sanyal and Amitabha Sinha

Chapter 8   145
Application of UAV Remote Sensing in Monitoring Banana
Fusarium Wilt
by Huichun Ye, Wenjiang Huang, Shanyu Huang, Chaojia Nie,
Jiawei Guo and Bei Cui

Chapter 9   161
Satellite Control System: Part II - Control Modes, Power, Interface,
and Testing
by Yuri V. Kim

Preface

This book presents the most recent environmental studies on remote sensing applications. Different sensors that start in both active and passive sensors are used in remote sensing applications. While passive remote sensing simply measures the electromagnetic radiation reflected from the target, active remote sensing transmits and measures electromagnetic radiation that is both emitted and reflected. Wide sensor coverage starts with satellite platforms and extends to unmanned aircraft vehicles (UAVs). What distinguishes a drone from a UAV? An unmanned aircraft or vessel that is piloted remotely or automatically is referred to as a drone. A UAV is a flying object without a pilot.

Remote sensing is a tool used for gathering target information without any physical/direct contact with the Earth's surface. It is a widely used science for the identification and mapping of the various objects/materials present on the Earth's crust. The electromagnetic wavelength ranges from 0.38 μm to 100 cm, which is visible to the microwave region and is utilized for capturing the information from the Earth's surface along with different sensors to capture the electromagnetic spectrum's energy. This technique is useful for monitoring, protecting, and managing diverse natural resources and land cover.

This book covers a variety of applications, including crop monitoring, ice automatic mapping, carbon offset forests, and image processing.

Chapter 1 is an introductory chapter that presents the novel Marghany-Based Genetic Algorithm (MBGA) for automatically detecting ice covers in Synthetic Aperture Radar (SAR) data. The MBGA is designed as discrete steps in a modified genetic algorithm (GA).

Chapter 2 highlights recent advances in SAR systems and their applications in crop growth monitoring. It provides an overview of recent advancements in SAR systems, a summary of SAR information sources, the applications in crop monitoring including crop classification, crop parameter estimation, and change detection, and perspectives for future application development.

Chapter 3 demonstrates a series of optical satellite data, specifically LANDSAT 8, in tracking the carbon offset identification in Malaysian forests. It demonstrates an excellent method for tracking and monitoring deforestation in Malaysia.

Chapter 4 provides an overview of the principles and applications of optical remote sensing in planetary science. It briefly introduces the planetary space environment before discussing the principles of optical remote sensing and planetary optical radiations. The chapter ends with a discussion of current and future optical remote sensing plans in China.

Chapter 5 scrutinizes a significant image enhancement tutorial. It considers both histogram modification and transform domain methods as well as hybrid methods.

Furthermore, it proposes a new hybrid algorithm for remote sensing image enhancement. Quality metrics such as Contrast Gain, Enhancement Measurement, Discrete Entropy, and Average Mean Brightness Error are also shown for objective comparison.

Chapter 6 serves as an excellent tool for identifying geological features. It shows how to distinguish between various minerals and lithologies using ASTER satellite data's short-wave infrared (SWIR) and thermal infrared (TIR) spectral bands. It also goes into detail about the value of integrated datasets from SWIR- and TIR-derived results and how to use them to demarcate different litho-units.

Chapter 7 presents transproc, a reconfigurable generic processor that can execute operations related to linear transformations like Fast Fourier Transform (FFT) and Forward Discrete Cosine Transform. (FDCT), or Forward Discrete Wavelet Transform (FDWT). A graph-theoretic lemma is used to find the applicability of such a processor to calculate the flow graph-related parallel operations found in these linear transformations. The architecture level design and processing element level design are also demonstrated in this chapter.

Chapter 8 introduces a technique based on a UAV. In this chapter, the multispectral images acquired by the UAV are exploited to establish a method to identify which banana regions were infected or uninfected with Fusarium wilt disease. The results suggest that UAV-based multispectral imagery with a red-edge band is effective to identify banana Fusarium wilt disease and that the CIRE had the best performance.

Finally, Chapter 9 considers various aspects of the optical and radiolocation sensing and imaging of the Earth's surface from space. These sensors (the payload), being placed on the satellite bus (platform), cannot be used for the satellite mission without its other subsystems. The mission itself, as well as the payload's successful operation and image-providing capability, essentially depends on Satellite Control System SCS and its performance.

I wish to convey my appreciation to the staff at IntechOpen, especially Editorial Project Manager Ms. Karmen Đaleta. Without their commitment and support, this book would not have been possible.

Dr. Maged Marghany
Professor,
Director,
Global Geoinformation, Sdn. Bhd.,
Kuala Lumpur, Malaysia

Introductory Chapter: Automatic Detection of Ice Covers in Airborne Radar Data Using Genetic Algorithm

Maged Marghany

1. Introduction

In the polar regions, where extreme weather poses significant challenges to field study, remote sensing from space is one of the most promising technologies being used to observe the environment. The climate, oceans, and terrestrial maritime ecology are all significantly impacted by sea ice, which is one of the most significant markers of changing climate in polar regions. As a consequence, several efforts have been made to keep track of the arctic sea ice.

The dual-core techniques, therefore, for tracking sea ice in the polar areas are the spaceborne radiometer (passive sensor) and scatterometer (active sensor). Ice cover volume, composition, and mobility were each determined using monitoring for the mission. The spaceborne microwave radiometer in particular provides the broadest time series of sea ice cover in the polar regions since 1979, demonstrating a reduction in the average sea ice cover of 0.53×10^6 km^2 per decade. In this sense, spaceborne synthetic aperture radar (SAR) is a superior option for tracking ice cover from a more comprehensive standpoint because of the major advantages of the high spatial and temporal resolution, polarimetric sensitivity, and configurable imagery modes. While spaceborne SAR can deliver sea ice cover measurements with good resolution at the dimension of 1 km and up to dozens of meters, the radiometer and scatterometer can deliver sea ice accumulation measurements over vast areas with a high resolution of 6.25 km to 12.5 km. In this understanding, Seasat, ERS-1/2, ENVISAT/ASAR, RADARSAT-1/2, TerraSAR-X/TanDEMX, and Sentinel-1 are some of the instances of spaceborne SARs that have proven to be efficient at tracking sea ice details, ice cover and amount, ice classification (such as ice floes, leads, and polynyas), ice movements and meander, icebergs, and ice-wave interactions. Despite techniques for charting sea ice cover and differentiating between ice cover and deep ocean from spaceborne SAR data having already historically been discussed, conventional both Antarctic and Arctic sea ice surveillance has not exploited these kinds of datasets [1–3].

Instinctually, the radar backscatter strength can indeed be used as the basis for the ice-water identification by spaceborne SAR data, because as backscatter of sea ice is usually higher than that of the deeper ocean. This is certainly relevant in the cross-polarization band, which is delicate to quantity scattering whereas the sea surface roughness commonly exhibits surface scattering. On the contrary, once incidence angles fluctuate, the radar backscatter of copolarization (vertical-vertical, VV, or horizontal-horizontal, HH) also varies significantly.

1

SAR cross-polarization signals, therefore, are demonstrated to be significantly extremely efficacious for identifying glaciers than copolarization SAR images, since the radar backscatter of the sea surface in cross-polarization is only marginally reliant on incidence angles and sea surface wind speed. The dual-polarization (HH and HV) SAR data of RADARSAT-2, Sentinel-1, and Gaofen-3 constitute the basis for certain newly proposed glacier classifier algorithms [1, 4].

In this view, a cornerstone of both Antarctic and Arctic zonal deviations is the volume of sea ice prevalent. Sea ice exhibits reasonably notable spatiotemporal variability in the marginal ice zone (MIZ), which suggests that satellite measurements of ice cover at a fine spatial resolution than operational radiometer and scatterometer merchandise are crucial due to the speeding up declining trend of ice covers and diminished ice thickness in both Antarctic and Arctic zones [1–3, 5–7].

The question is now: what are the main algorithms exploited in retrieving ice covers in the SAR data? Despite the similarities between ice cover and the open ocean's SAR radar backscatter in certain circumstances, their textural elements that are dependent on brightness might fluctuate. To distinguish ice cover and deep ocean, descriptors are employed as well in conjunction with radar backscatter strength. The gray-level cooccurrence matrix (GLCM) texture characteristics can accurately capture distinctive backscatter (or brightness) features that are distinct across varying forms of the glacier and open sea, according to numerous research on texture examinations of SAR imagery [8, 9]. The textural characteristics of energy, contrast, correlation, homogeneity, entropy, and moment may be useful for classifying ice, according to earlier research [8–10].

Recent algorithms are exploited in glacier classifications based on the deep learning machine. Particularly, the support vector machine (SVM), a well-liked binary classification deep learning technique, has been focused on the identification of ocean glacier water (hence shortened to ice water) from spaceborne SAR data [1, 9, 11]. For numerous spaceborne SAR data, pixel-based and region-based glacier categorization methods have been invented [1, 4, 10, 12].

The critical question is: what are the disadvantages of using SVM in ice classifications in SAR data? Since the SVM employs a machine learning approach, acquiring a well-labeled training set that can distinguish between ice extent and deep ocean in Image data is probably the hardest process. Various glacier sorts, like multiyear glaciers, extent and distorted first-year ice, young ice, and numerous more, as well as various open ocean varieties, such as calm and rough sea surfaces, ought to be included in the training set. It takes a considerable amount of time and effort to acquire training datasets, which are often chosen extensively by professionals.

The novelty of this chapter is to form a novel algorithm based on the genetic algorithm (GA) for the automatic detection of ice covers in the Convair-580 aircraft. In actuality, the Convair-580 research aircraft serves as a multifunctional flying laboratory that supports a wide variety of studies. The normalized radar cross-section (NRC) Convair-580 is outfitted with cutting-edge technology for detecting aircraft physical parameters and atmospheric state (temperature, pressure, humidity, and three-dimensional wind).

2. Convair-580 data acquisition

Spectral ranges of 1000 nm to 2450 nm are covered by the 160-channel Short Wave Infrared (SWIR) hyperspectral imaging system. These data include a completely polarimetric dual-frequency (W and X-band) Doppler radar system. The Convair (**Figure 1**) will incorporate this NRC Airborne W and X-band radar system (NAWX) by January 2006. The thermal microwave emission from the surface and

Figure 1.
Convair-580 research aircraft is used in this study.

atmosphere is measured by the Airborne Multichannel Microwave Radiometer (AMMR) and is expressed in degrees Kelvin of brightness temperature. In the 1970s, the uplooking radiometer at 21 and 37 GHz, a part of AMMR, was created to monitor precipitation from an aircraft. The entire AMMR assembly operates between 10 and 92 GHz. Over the past three decades, a variety of aircraft has used the 21/37 GHz unit.

3. Marghany-based genetic algorithm (MBGA) for automatic detection of ice covers in Airborne SAR data

This section introduces a developed genetic algorithm for ice cover automatic detection in any SAR data products. This algorithm is named as Marghany-Based Genetic Algorithm (MBGA) (**Figure 2**), which was adapted from the previous work of Marghany [15].

```
FUNCTION _MBGA
1. Initialize Number_of_ice pixel generations=0
```

2. Generate $f(\beta) = [f_1(\beta), f_2(\beta), ..., f_k(\beta)]^T$

```
//-------Run MBGA----------------
```

3. While Number_of_generations< $\tau = 0.5 \, [Max \, f(P^j) + Min \, f(P^j)]$

3.1 $f(P^j) = [\sum_{i=1}^{N} \frac{\beta_i}{N} \sum_{i=1}^{K} |P_i^j - \beta_i|]^{-1}$

3.2 Select pairs of individuals from the population

3.3 Crossover $f(P_i^j) = |\beta_i - P_i^j|$

```
 3.4 Increment Number_of_generations
 End //---- end while--------
END
```

Figure 2.
Pseudo-code depicting MBGA algorithm.

Let K be the total backscattered energy in the Convair-580 Aircraft data, and $[(\beta_1), (\beta_2), ..., (\beta_n)]$ be the sum of all backscattered bright pixels. Since genetic algorithms begin with the population initializing phase, K is composed of genes that represent the backscatter of bright pixels and their surroundings [13]. In this view, GA might be considered as having followed Sivanandam and Deepa [14] as:
Minimize

$$f(\beta) = [f_1(\beta), f_2(\beta), ..., f_k(\beta)]^T \tag{1}$$

Formula (1) demonstrates that $f_i(\beta)$ presents the *i-th* pixel backscatter β discrepancies in Convair-580 data, which signifies the *i-th* and *j-th* restraints backscatter energy in raw direction and column direction, respectively. Consequently, a fitness function is nominated to regulate the resemblance of separately backscatter energy associated with ice covers in the Convair-580 image. In this view, the backscatter of ice covers is signified by where $i = 1,2,3, ..., K$ and the initial population where $j = 1,2,3, ..., N$ and $i = 1,2,3 ..., K$. Consequently, the fitness value of every separated population of the radar backscatter energy mathematical expressed as:

$$f(P^j) = \left[\sum_{i=1}^{N} \frac{\beta_i}{N} \sum_{i=1}^{K} \left|P_i^j - \beta_i\right|\right]^{-1} \quad j = 1, ..., N. \tag{2}$$

In this perspective, N and K represent the number of the population involved in the fitness mechanism. In most scenarios, Convair-580 aircraft data exploits Eq. (2) to assess the level of similarity of phase corresponds associated with ice covers. Population sizes have been generated before this computation. Consequently, let us assume that P_i^j is a gene which corresponds to backscatter energy fluctuation through SAR data. Accordingly, P_i^j are chosen at random to illustrate the backscatter changes of the ice cover pixels as well as their surroundings. Additionally, i diverges from 1 to K and j fluctuates from 1 to N where N is the population dimensions.

In this understanding, let us consider the best fitness selection of individuals' backscatter energy $f(P^j)$ from the population P_i^j. Therefore, the maximum values of fitness of the population $Max\, f(P^j)$ and the minimum values of fitness of the population $Min\, f(P^j)$ are exploited to compute the threshold value τ, which is casted as:

$$\tau = 0.5 \left[Max\, f(P^j) + Min\, f(P^j)\right] \tag{3}$$

The empirical formula (3) is employed as a phase in the selection process to establish the population's maximum and lowest fitness levels, respectively. In GA algorithms, this is regarded as the population generation phase for brightness patches in SAR data [15, 16].

The reproduction stage of a genetic algorithm, which incorporates crossover and mutation processes on the backscatter population P_i^j in Convair-580 data, is primarily responsible for its operation. The crossover operator shapes the P_i^j to converge around solutions with high fitness. Therefore, the convergence occurs more rapidly the closer the crossover probability is to 1. The chromosomes exchange genes during the crossover process. Depending on the local fitness value, each gene can be formed by:

$$f\left(P_i^j\right) = \left|\beta_i - P_i^j\right| \tag{4}$$

In that circumstance, the crossover between two individuals process converts all individual populations of the first parent that have a local fitness $f\left(P^j_{av}\right)$ that is higher than the average local fitness and replaces the remaining genes with the matching ones from the second parent. Thus, the inclusion conditions characterize the average local fitness:

$$f\left(P^j_{av}\right) = \frac{1}{K}\sum_{i=1}^{K}\left|\beta_i - P^j_i\right| \tag{5}$$

The phenomenon of remarkable probability in the evolution process is thus denoted by the mutation operator. There is a potential that certain crucial genetic data about the chosen population will indeed be lost throughout the reproduction process. Hence, the mutation operator brings additional genetic sequences to the genetic diversity.

4. Results and discussion

Figure 3a exhibitions a Convair-580 image with brightness patches designating the glacier zone. In contrast to the surroundings, **Figure 3b** reveals that the glacier patches seemed to have the maximum backscatter (−8 dB). The smallest backscatter, measuring −55 dB, is scattered among dark pixels that may represent calm water or a low wind zone.

The crossover process for the Convair-580 image is shown in **Figure 4** and involved 10 individuals. Positive bright spots in these 10 individuals represent the pixels that make up the ice cover, while negative dark patches represent the pixels that make up the surroundings, particularly still water or melting ice. Every cell is then compared to its counterpart in the other cells to determine if it is positive or negative. This study supports the work of Marghany [13] and Ninnis et al., [17].

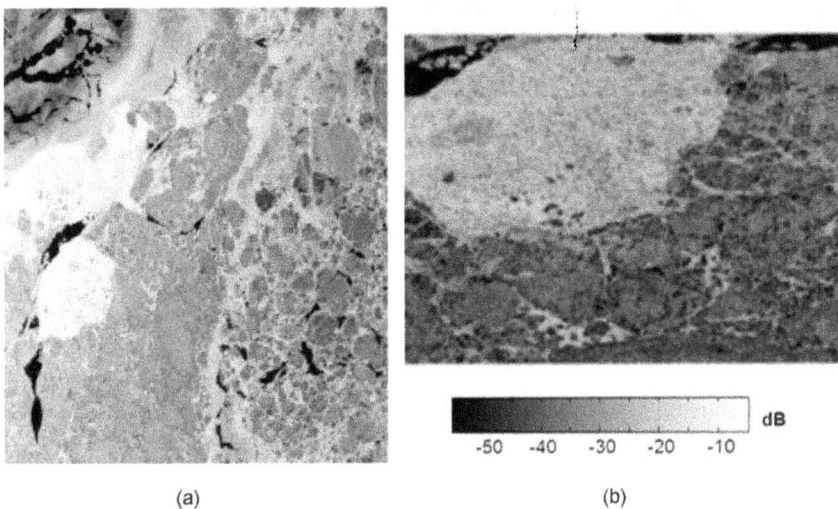

(a)   (b)

Figure 3.
Data from a Convair-580 aircraft, shown as (a) composite bands of VV, HH, and HV and (b) backscatter variation.

Figure 4.
First individual in the crossover phase.

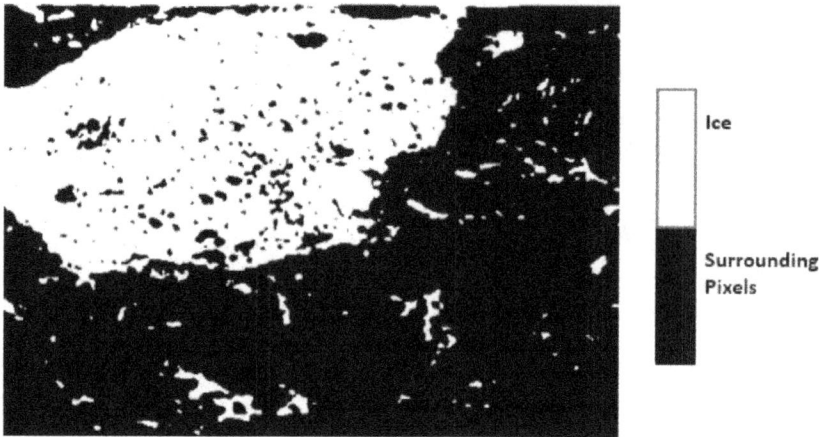

Figure 5.
Automatic detection of ice covers in SAR data using MBGA.

Figure 5, consequently, demonstrates how the genetic algorithm can effectively isolate ice pixels from their surroundings. In other words, heavy ice covers, ice boundaries, and edges are all colored white, whereas calm water and melting ice pixels are all colored black (**Figure 2a**). This research is similar to Marghany's [13] previous work on using GA for object detection.

The significant question that arises is: what is the appropriate radar polarimetry for ice cover imaging? In this sense, **Figure 6** demonstrates that the HH band has lower standard errors of 17% than other bands VV and HV; respectively. On the contrary, HV has the highest standard error of 79%.

Consequently, the Marghany-Based Genetic Algorithm (MBGA) can also discriminate between different sorts of ice covers such as leads, young ice, and floes (**Figure 7**) in the HH band owing to its long tilt modulation as compared to VV and HV bands, respectively. In this sense, the receiver operating characteristic (ROC)

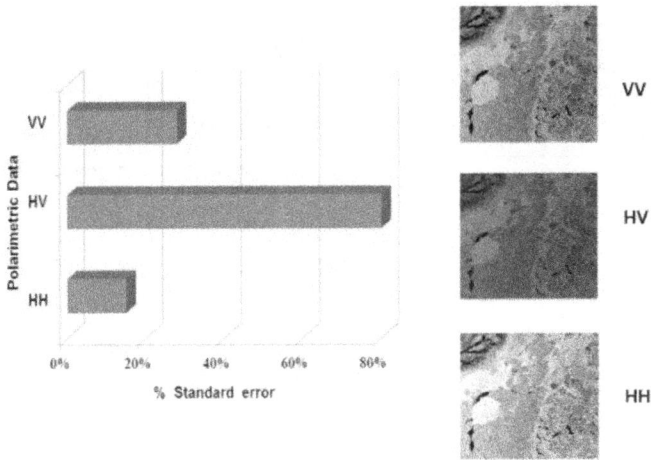

Figure 6.
Standard errors among different polarized bands.

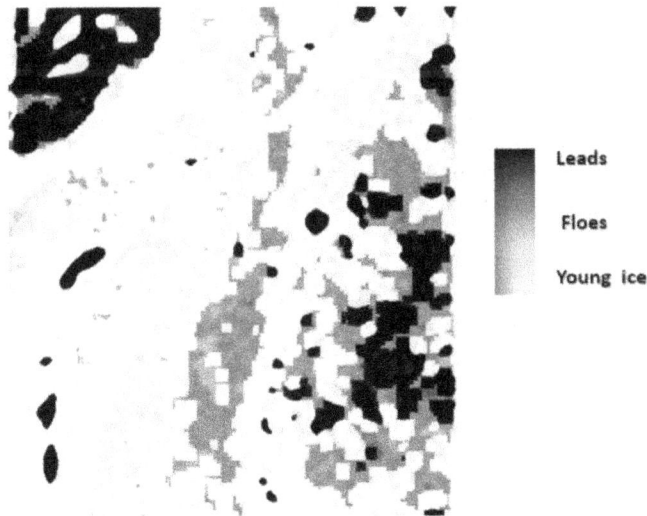

Figure 7.
Automatic detection of different sorts of ice covers using MBGA.

curve in **Figure 8** shows a significant difference in discrimination power between pixels representing leads, first-year ice, and floes with a probability $p < 0.05$.

The crossover procedure is what grants the MBGA its power. Each crossover process creates a new population. As a result, the fitness function scrutinizes multiple individual populations and incorporates them into subsequent populations. As a consequence, new populations are constantly generated based on the differences between two successive fitness values. Furthermore, the crossover procedure produces a more refined ice cover pattern by despeckling and preserving the morphology of the features of the ice cover groups through the fitness function used to implement the ice covers in different pixel classes. Indeed, the fitness function chooses a morphological pattern for the ice covers that is similar to the ice cover sorts that are recommended.

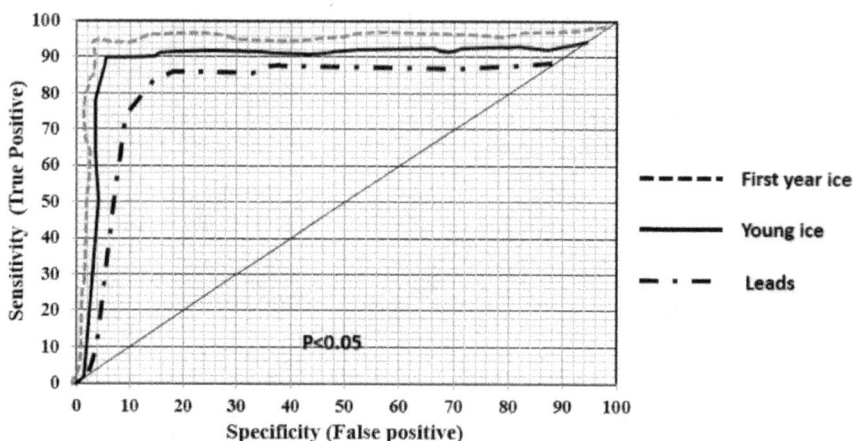

Figure 8.
ROC for ice sort discriminations using genetic algorithm (GA).

5. Conclusion

This chapter introduces the Marghany-Based Genetic Algorithm (MBGA), a novel algorithm for automatically detecting ice covers in SAR data. MBGA is thus designed as discrete steps of a modified genetic algorithm (GA). Individual backscatter fluctuation in SAR data is used as the primary source, which is generated by sequences of genetic algorithm procedures. In this scenario, MBGA outperforms other bands in terms of automatic detection of ice covers within the HH band, with the lowest standard errors of 17%. As a consequence, MBGA can automatically distinguish between different types of ice covers, such as leads, floes, and young ice. This is demonstrated using ROC, which indicates excellent discrimination of ice cover of different kinds with $p < 0.05$.

Author details

Maged Marghany
Global Geoinformation Sdn.Bhd., Masakiara Residences, Kuala Lumpur, Malaysia

*Address all correspondence to: magedupm@hotmail.com

IntechOpen

References

[1] Li XM, Sun Y, Zhang Q. Extraction of sea ice cover by Sentinel-1 SAR based on support vector machine with unsupervised generation of training data. IEEE Transactions on Geoscience and Remote Sensing. 2020;**59**(4): 3040-3053

[2] Kwok R, Schweiger A, Rothrock DA, Pang S, Kottmeier C. Sea ice motion from satellite passive microwave imagery assessed with ERS SAR and buoy motions. Journal of Geophysical Research: Oceans. 1998;**103**(C4): 8191-8214

[3] Walker NP, Partington KC, Van Woert ML, Street TL. Arctic Sea ice type and concentration mapping using passive and active microwave sensors. IEEE Transactions on Geoscience and Remote Sensing. 2006;**44**(12): 3574-3584

[4] Minwei Z, Xiaoming L, Yongzheng R. The method study on automatic sea ice detection with GaoFen-3 synthetic aperture radar data in polar regions. 海洋学报. 2018;**40**(9):113-124

[5] Zakhvatkina N, Korosov A, Muckenhuber S, Sandven S, Babiker M. Operational algorithm for ice–water classification on dual-polarized RADARSAT-2 images. The Cryosphere. 2017;**11**(1):33-46

[6] Zakharov I, Power D, Howell M, Warren S. Improved detection of icebergs in sea ice with RADARSAT-2 polarimetric data. In: 2017 IEEE International Geoscience and Remote Sensing Symposium (IGARSS). Fort Worth, TX, USA: IEEE; 2017. pp. 2294-2297

[7] Silva TA, Bigg GR. Computer-based identification and tracking of Antarctic icebergs in SAR images. Remote Sensing of Environment. 2005;**94**(3): 287-297

[8] Clausi DA. Comparison and fusion of co-occurrence, Gabor and MRF texture features for classification of SAR Sea-ice imagery. Atmosphere-Ocean. 2001; **39**(3):183-194

[9] Clausi DA, Zhao Y. Grey level co-occurrence integrated algorithm (GLCIA): A superior computational method to rapidly determine co-occurrence probability texture features. Computers & Geosciences. 2003;**29**(7): 837-850

[10] Zakhvatkina NY, Alexandrov VY, Johannessen OM, Sandven S, Frolov IY. Classification of sea ice types in ENVISAT synthetic aperture radar images. IEEE Transactions on Geoscience and Remote Sensing. 2012; **51**(5):2587-2600

[11] Marghany M. Flock 1 data multi-objective evolutionary algorithm for turbulent flow detection. In: CD of 36th Asian Conf Remote Sensing: Fostering Resilient Growth in Asia. ACRS; 2015. pp. 1-6

[12] Tan W, Li J, Xu L, Chapman MA. Semiautomated segmentation of Sentinel-1 SAR imagery for mapping sea ice in Labrador coast. IEEE Journal of Selected Topics in Applied Earth Observations and Remote Sensing. 2018;**11**(5):1419-1432

[13] Marghany M. Genetic algorithm for oil spill automatic detection from ENVISAT satellite data. In: International Conference on Computational Science and its Applications. Berlin, Heidelberg: Springer; 2013. pp. 587-598

[14] Sivanandam SN, Deepa SN. Introduction to Genetic Algorithms, by springer berlin heidelberg new york. 2008

[15] Marghany M. Utilization of a genetic algorithm for the automatic detection of

oil spill from RADARSAT-2 SAR
satellite data. Marine Pollution Bulletin.
2014;**89**(1-2):20-29

[16] Ressel R, Frost A, Lehner S. A neural
network-based classification for sea ice
types on X-band SAR images. IEEE
Journal of Selected Topics in Applied
Earth Observations and Remote
Sensing. 2015;**8**(7):3672-3680

[17] Ninnis RM, Emery WJ, Collins MJ.
Automated extraction of pack ice
motion from advanced very high
resolution radiometer imagery. Journal
of Geophysical Research: Oceans. 1986;
91(C9):10725-10734

Recent Advancement of Synthetic Aperture Radar (SAR) Systems and Their Applications to Crop Growth Monitoring

Jiali Shang, Jiangui Liu, Zhongxin Chen,
Heather McNairn and Andrew Davidson

Abstract

Synthetic aperture radars (SARs) propagate and measure the scattering of energy at microwave frequencies. These wavelengths are sensitive to the dielectric properties and structural characteristics of targets, and less affected by weather conditions than sensors that operate in optical wavelengths. Given these advantages, SARs are appealing for use in operational crop growth monitoring. Engineering advancements in SAR technologies, new processing algorithms, and the availability of open-access SAR data, have led to the recent acceleration in the uptake of this technology to map and monitor Earth systems. The exploitation of SAR is now demonstrated in a wide range of operational land applications, including the mapping and monitoring of agricultural ecosystems. This chapter provides an overview of—(1) recent advancements in SAR systems; (2) a summary of SAR information sources, followed by the applications in crop monitoring including crop classification, crop parameter estimation, and change detection; and (3) summary and perspectives for future application development.

Keywords: synthetic aperture radar (SAR), crop growth monitoring, crop parameter estimation, change detection, classification

1. Introduction

Agricultural ecosystems are highly dynamic and usually display apparent seasonal phenological patterns that are strongly dependent on local management practices. The timely and frequent determination of indicators of crop development and productivity, including phenological stage and biophysical parameters such as leaf (or plant) area index and above-ground biomass, is critical for supporting land management decision making in near-real-time. Synthetic aperture radars (SARs) are active systems that provide their own source of energy to illuminate ground targets in the microwave domain. Because the Earth's atmosphere is largely transparent to microwaves, SAR sensors can be operated day or night and under almost at all weather conditions to acquire high-resolution earth observation data. Given that many regions of the world experience frequent cloud cover, SAR has become an essential remote sensing tool for the operational monitoring of agricultural production systems around the world.

Radar backscattering is highly sensitive to the structural (roughness, orientation, and spatial distribution of scattering components) and the dielectric properties of targets. The backscattering intensity is also strongly related to the transmitted microwave frequency, incident angle and the transmitted and received polarizations. Several microwave scattering models have been developed to relate backscattering to target properties and radar acquisition parameters. Examples of theoretical microwave models include the Integral Equation Model (IEM) and the MIMICS (Michigan Microwave Canopy Scattering) model [1–3]. Semi-empirical models maintain some theoretical basis but use empirical data to simplify the mathematical relationships between scattering and target properties as well as sensor parameters. Examples of this approach to modeling include the Water Cloud Model (WCM) used to characterize SAR response from vegetation and soil [4], as well as the Oh [5] and Dubois [6] models that relate soil properties to radar backscattering. Two simplified scenarios, based on which radar backscattering models have been developed, are shown in **Figure 1**. The first simplifies vegetation canopy as a layer of scattering elements uniformly distributed above the soil surface, and radar backscattering from the soil is modeled by a two-way attenuation through the canopy (left). The second takes into consideration of canopy geometric structure, and models three backscattering components, surface scattering from plant or soil, double-bounce scattering from plant and soil (plant-soil and soil-plant), and multiple scattering by the plant-soil mix (volume scattering) (right).

Radar backscattering models have been used for estimation of crop parameters such as Leaf Area Index (LAI), canopy water content, and biomass [7–10], and soil parameters such as soil moisture and surface roughness [11, 12], using SAR data acquired at different incident angles, frequencies, and/or polarizations. Fully polarimetric (or quad-pol) SAR systems measure the complete complex scattering from a target. Microwaves are transmitted and received in two orthogonal polarizations and the phase is preserved during processing. With the complete scattering matrix, quad-pol data can be analyzed to provide polarimetric features and the signal can be decomposed using coherent (e.g., Pauli and Cameron) or incoherent (e.g., Freeman-Durden and Cloude-Pottier) techniques [13, 14]. Variables derived through polarimetric decomposition can be used both in classification [15, 16] and parameter estimation, such as crop phenology or soil moisture [17, 18]. Time-series SAR data have also been used for the detection of crop seeding and harvest using change detection approaches [19–21].

A few satellite SAR constellations have been launched during the past few years, and more small satellite SAR constellations will be continuously developed in near future. The increasing availability of a large amount of SAR data, in companion

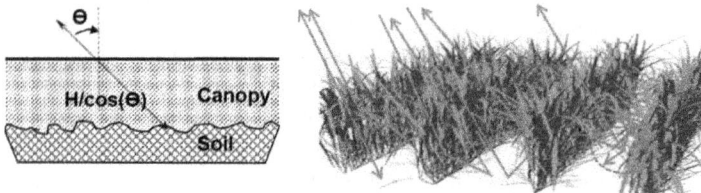

Figure 1.
Simplified scenarios for modeling radar backscattering from vegetation canopy. Left: vegetation as a water cloud, and backscattering is modeled by a two-way attenuation through a canopy with a path length H/cos(Θ). Right: vegetation as a 3-D scattering medium inducing three scattering mechanisms, surface scattering, double-bounce, and volume scattering.

with big data analytics, provides an unprecedented opportunity for effective and operational monitoring of agricultural ecosystems. However, the effective use of SAR data requires a full understanding of how the information it provides relates to agricultural targets. The objectives of this article are to summarize the path of SAR system development and to review the information sources of SAR data, the applications in agricultural ecosystem monitoring with a focus on crop classification, crop parameter estimation, and change detection using dense time-series data.

2. Advances in synthetic aperture radar (SAR)

Early studies in radar remote sensing applications in agriculture relied extensively on ground-based microwave scatterometers [22–27]. The portability of scatterometers allows them to be rapidly deployed to agricultural test sites to collect temporally dense data at different frequencies, polarizations, and incidence angles. Experiments using scatterometers have been critical for developing an understanding of how microwaves interact with soils and crops, and the development and testing of microwave models [28]. However, despite the important contributions of such research, scatterometer data are geographically limited to smaller test plots.

The deployment of SAR on aircraft and satellite platforms provides data at field and sub-field scales over much broader geographic extents. Airborne SAR campaigns, such as the NASA/JPL AIRSAR and UAVSAR, Canadian Convair-580 C/X SAR, and German DLR E-SAR/F-SAR, have served as theoretical testbeds to develop applications pre-launch of space-based SARs. Space-based SAR observations from the Shuttle Imaging Radar (SIR) missions, in particular the SIR-C/X SAR missions in 1994, provided imaging opportunities from a space platform and delivered data in different frequencies and polarizations.

Systematic acquisitions from SAR satellites began with the launch of ESA's ERS-1 satellite in 1991. Several other space agencies followed, launching SAR satellites operating at different frequencies, and with different capacities to select imaging modes at a variety of spatial resolutions, swath widths, polarization, and incident angles (**Table 1**). RADARSAT-2, for example, supports the acquisition of data at single, dual, or quad polarization, at different spatial resolutions, and various incident angles. However, while the capability of each of these space-based systems was extensive, they demanded large and heavy payloads. For example, the mass of the Canadian C-band RADARSAT-1 and -2 satellites, and the ESA Sentinel-1A and 1B satellites each exceeded two tons at launch. More recently, technological developments that include standard electronic components and semiconductor materials (GaN) [29, 30] make it possible to produce compact SAR sensors in a shorter amount of time, and at a relatively low cost. These advancements have led to commercial investments in microsatellite constellations of space-borne SARs. For instance, PredaSAR plans to launch a constellation of 48 satellites equipped with a large swath C-band or a high-power X-band sensor. The Japanese QPS Institute is developing an X-band constellation that will eventually comprise 36 micro-satellites. These SAR sensors are typically small (<500 kg), but are more limited in the diversity of imaging modes, typically operating in only single or dual polarizations.

For reference, a non-exclusive list of SAR systems that are of interest to agricultural applications is given in **Table 1**. Over the past 15 years or so, the general trend of governments and space agencies has been to focus on larger wide-swath SARs whose data are free and open (or partially open) to the public. In comparison, the commercial SAR satellite ecosystems have focused on constellations of smaller satellites providing, for a fee, access to data at finer spatial and temporal resolutions. Data from such constellations may provide the near-continuous monitoring of land surfaces.

Platform	Country/ organization	SAR system	Frequency	Mode	Active years	Note
Airborne	Canada	Convair-580	X, C	Polarimetric	1986–present	
	USA/NASA	AirSAR	C, L, P	Polarimetric	1988–2004	
	USA/NASA	UAVSAR	Ka, L, P	Polarimetric	2007–present	
	German/ DLR	E-SAR/F-SAR	X, C, S, L, P	Polarimetric	1988–present	
	USA.JPL	SIR-C/X	C, X	Polarimetric	1994	
Large satellites	ESA	ERS-1/2	C	VV	1991–2011	
	ESA	ASAR	C	Various	2002–2012	
	Japan/ NASDA	JERS-1/2	L	HH	1992–1998	
	Canada	RADARSAT-1	C	HH	1995–2013	
	Canada	RADARSAT-2	C	Various	2007–present	
	Canada	RCM	C	Various	2019–present	3 satellites
	German	TerraSAR-X	X	Single or dual	2007	
	Argentina	SAOCOM	L	Polarimetric	2018–present	2 satellites
	ESA	Sentinel-1	X	Single or dual	2014–present	4
	Italy	COSMO-SkyMed	X	Various	2007–2010	COSMO: 4
					2019–present	CSG: 2
	Japan/JAXA	ALOS-PALSAR	L	Various	2006–present	4
	USA/India	NISAR	L, S	Polarimetric	2023–present	
Micro-satellites	Finland	ICEYE-X	X	VV	2018–present	18
	Japan/ Synspective	StriX	X	VV	2020–present	30
	Japan/QPS	QPS-SAR	X	Circular	2019–present	36
	USA	Capella	X	HH	2018–present	36
	USA	PredarSAR	C, X		—	48
	USA	Umbra-SAR	X		2021–present	12

Abbreviations/websites: DLR: German Aerospace Center, ESA: European Space Agency, JAXA: Japan Aerospace Exploration Agency, JPL: Jet Propulsion Laboratory, NASA: National Aeronautics and Space Administration (USA), NASDA: National Space Development Agency of Japan, PredSAR: www.predasar.com, QPS: Institute for Q-shu Pioneers of Space, Inc.; https://i-qps.net/, Synspective: https://synspective.com/, AIRSAR: Airborne Synthetic Aperture Radar, ALOS-PALSAR: Phased Array type L-band Synthetic Aperture Radar; https://www.eorc.jaxa.jp/ ALOS/en/about/palsar.htm, ASAR: Advanced Synthetic Aperture Radar, Capella: https://www.capellaspace.com/, Convair-580: https://open.canada.ca/data/en/dataset/838aa171-efa0-4951-9fad-37f9d99346ec?=undefined&w bdisable=true, COSMO-SkyMed: Constellation of Small Satellites for Mediterranean basin Observation; https:// earth.esa.int/web/eoportal/satellite-missions/c-missions/cosmo-skymed, ERS-1/2: European Remote-Sensing Satellite, E-SAR/F-SAR: https://www.dlr.de/hr/en/desktopdefault.aspx/tabid-2326/3776_read-5691, ICEYE: https://www. iceye.com, JERS-1/2: Japanese Earth Resources Satellite, NISAR: NASA-ISRO Synthetic Aperture Radar; https:// nisar.jpl.nasa.gov, PredarSAR: https://www.predasar.com/, RCM: Radarsat Constellation Mission, SAOCOM: https://saocom.veng.com.ar/en/, Sentinel-1: https://sentinel.esa.int/web/sentinel/missions/sentinel-1, SIR-C/X: Shuttle Imaging Radar, StriX: https://synspective.com/satellite/satellite-strix/, TerraSAR: https://www.dlr.de/ content/en/articles/missions-projects/terrasar-x/terrasar-x-earth-observation-satellite.html, UAVSAR: Uninhabited Aerial Vehicle Synthetic Aperture Radar; https://uavsar.jpl.nasa.gov, Umbra-SAR: https://umbra.space/.

Table 1.
List of SAR systems.

3. SAR applications to crop growth monitoring

3.1 Sources of information

3.1.1 Multi-temporal acquisitions

Crop growth dynamics are characterized by structural or moisture changes, which can be captured by time-series SAR data for the detection and mapping of phenological development stages [17, 31–34]. Because different crops have different growth dynamics, time-series SAR is also useful for crop classification [35–38]. Time-series optical and SAR data have been used to derive phenological metrics for phenology-based crop classification [35, 39, 40]. Using a dense stack of Sentinel-1 SAR data, Bargiel [35] proposed a crop classification scheme using phenological sequence patterns (PSP), which outperformed the Random Forest and the Maximum Likelihood classifiers for cereal crops. Phenology-based classifiers can be more generic than conventional classifiers, and may be more resilient to differences in management practices and growth conditions because they take crop-specific growth dynamics into consideration [35, 39].

3.1.2 Polarizations and polarimetric decomposition

Important target information is also revealed by different SAR polarizations. Polarizations that interact more strongly with plant volume will likely be more useful for crop parameter estimation and for discriminating different crops. For example, VV performs well in characterizing vertical vegetation structure, and VH is sensitive to multiple scattering events in the canopy [41], and thus their use in combination provides better classification capabilities in most cases. HH polarization is found to be inferior in many cases for these specific applications [28, 42]; however, it is more sensitive to the structural variation of rice, and thus useful for mapping this crop [43].

SAR backscatter intensity and other polarimetric parameters can be derived from fully polarimetric SAR using coherent or incoherent target decomposition methods, as summarized in Cloude and Pottier [13], Touzi et al. [14], and Lee and Pottier [44]. Simple or canonical targets—such as dipoles, diplanes, or cylinders—show higher coherence than distributed targets—such as rough soil surfaces or vegetation—where random scattering occurs. Criteria to determine coherent and incoherent targets are provided by Touzi and Charbonneau [45], using the maximum symmetric component derived from the Cameron decomposition. Coherent target decomposition is applied to express the complex scattering matrix as linear combinations of a set of simpler and independent bases, each representing certain physical scattering mechanisms. Examples include the Pauli decomposition, the Krogager decomposition, and the Cameron decomposition. Incoherent decomposition methods are used when a pixel contains distributed targets, and express the second-order statistics of coherency or covariance matrices with a combination of simpler components. Examples include the Freeman-Durden, Huynen, and Cloude-Pottier decompositions. For satellite SAR sensors, the power and pulse repetition required to operate a fully polarimetric system limits swath widths, and thus hybrid architectures, such as compact polarimetric systems, have been proposed [46]. Compact polarimetry offers a partial solution by transmitting a single circularly polarized wave and receiving two orthogonal waves coherently [46–48]. Methods for compact polarimetric data decomposition have also been developed and summarized in Charbonneau et al. [47], Cloude et al. [49], and Ponnurangam and Rao [50].

3.1.3 Frequencies

The distribution and orientation of plant components and their sizes relative to SAR wavelengths vary over the growing season and from crop to crop. Microwave scattering occurs when the SAR wavelength is similar to or smaller than the size of canopy components. SAR frequency also influences the penetration depth of microwave radiation into crop canopies. Lower frequencies (e.g., L-band) penetrate deeper into the canopy than higher frequencies (e.g., C-/X-band). The optimal depth of penetration, and the matching of wavelength to the size of plant components, vary from crop to crop and throughout the crop development cycle. As a result, the selection of a single best frequency for SAR is challenging. Higher frequency SAR is better for classifying low biomass canopies, while lower frequency SAR is more useful for identifying high biomass vegetation [42]. The integration of data at different frequencies brings enriched information for crop classification and has thus been widely recommended [42, 51–57]. However, implementing a multi-frequency approach is challenging due to limitations in the availability of data from sensors at different frequencies, especially for operational applications. Temporal signatures created by different frequencies have also been exploited for crop area mapping using dense time series of SAR data. Kraatz et al. [58] used the temporal coefficient of variation of the VH polarization from both Sentinel-1 C-Band and PALSAR L-band data, and an optimal threshold, to discriminate crops from non-crops in western Canada. A higher mapping accuracy was achieved using C-band data (84%) than L-band data (74%), though performance varied among different cover types. Here, L-band performed poorly for soybean and some non-crop types (urban, grassland, and pasture), while C-band was relatively poor for corn, urban, and pasture. A time series of data from both frequencies would likely have improved these accuracies.

3.1.4 Incident angles

The variation of radar backscattering with incident angle is another important consideration for mapping agricultural landscapes with SAR. This is reflected in vegetation backscattering models, such as the MIMICS model [3] and the Karam-Fung model [59]. These models, developed for forest and adapted for crops [43, 60], require incident angle as an explicit parameter. For example, Prevot et al. [61] showed that using a simple parametrization of the angular effect of soil roughness in the Water Cloud Model [4], the vegetation water content can be estimated satisfactorily from C- and X-band SAR data acquired at two different incident angles, for example, 20° and 40°. Various studies have demonstrated the impacts of incident angle on land cover classification. Poirier et al. [62] studied the impacts of incident angle on classification performance by acquiring C-band data near-coincident at two different angles (30° and 53°) with the Convair-580 airborne system. Results showed that SAR data collected at the larger incident angle interacted more with the upper canopy, delivering an improved classification. Kothapalli Venkata et al. [63] conducted a study to assess the separation of corn from other land cover types (wheat, fallow, water, and urban) using multi-incident angles (28°, 42°, and 52°) C-band hybrid polarimetric data acquired over 3 days by RISAT. The study showed that corn can be discriminated from other crop types using volume and double-bounce scattering at both 28° and 42°, and using odd bounce and volume scattering combinations at 52°. Xu et al. [64] acquired RADARSAT-2 data at three different incident angles and showed that multi-angle SAR improved the classification accuracy of some land cover types (though it should be noted that the images were acquired at different times during 1 month, confounding the effects of the time of acquisition and change in incident angle). In summary, SAR data acquired at different incident angles contribute to target information extraction.

3.2 Crop type classifications

The classification of land covers and crop types is one of the earliest applications of SAR in agriculture. In the broadest sense, crop classification from SAR involves the implementation of automated techniques to sort image data into one of a finite number of crop classes based on their backscatter characteristics. Crop classification is an important agricultural application because it can be used to derive the area seeded to individual crops, and to predict or forecast food production if crop growth conditions are incorporated. Obtaining this information requires detailed, routine, and frequent mapping of croplands with sufficiently high accuracy. SAR has been shown to be particularly useful for the operational monitoring of crop dynamics in agricultural ecosystems.

3.2.1 Classification algorithms

A broad array of approaches for classifying satellite images have been developed in the past few decades. Until recently, the Maximum Likelihood (ML) classifier was the most widely used method for the supervised classification of remote sensing data [65–67], mostly due to its simplicity in implementation. While this approach has been widely applied in different studies for satellite image classification of agricultural regions [68–71], limitations associated with the ML approach mean that alternative supervised classification techniques are more preferable. Of these new methods, artificial neural networks (ANN) [72–75], support vector machines (SVM) [76–79], Decision Trees (DT) and ensembles of classification trees such as Random Forest (RF) [80–84] have all shown great promise.

A detailed comparison of classification methods is beyond the scope of this article, and indeed, would only provide limited insight into the best classification approaches for SAR-based crop type mapping. This is because the success of crop classification procedures is as much—if not more—dependent on the quality of the ground (in situ) observations used for training and validating the classification, than the actual algorithm chosen to do the classification. Instead, we direct readers to a comprehensive synthesis of this body of work provided by Khatami et al. [85], who conducted a statistical meta-analysis of research on land-cover classification. This study was conducted to provide coherent guidance on the relative performance of different classification processes for generating land cover products and showed that the highest mapping accuracies were provided by implementations of SVMs, ANNs, and RF. While it is important to note that these results are not necessarily predictive of the relative performance of any specific classifier in any specific application (due to the unique features of that application), they do provide an insight into how each classification algorithm may perform under various circumstances [85].

In addition to the general classifiers presented above, two other classification schemes have been developed specifically for SAR data. These are classifications based on the Cloude-Pottier decomposition and classification based on the complex Wishart distribution. The Cloude-Pottier decomposition [15] produces three parameters—entropy (H), anisotropy (A), and the alpha angle (α). Entropy is a metric of the degree of randomness of scattering from within the resolution cell, anisotropy is an indicator of the presence of secondary or tertiary scattering mechanisms, and the alpha angle represents the dominant scattering mechanism. A classification scheme was developed to divide the H-α space into eight possible scattering zones, from which land cover classifications can be performed. The advantage of this classification scheme is the improved understanding of SAR signal scattering mechanisms where there is less *a priori* knowledge about the scene. This approach has been used in supervised and unsupervised classification algorithms for land cover classification [86–90].

SAR data are typically multi-look processed for speckle noise reduction. The covariance matrix of the multi-look processed SAR data follows a multivariate complex Wishart distribution. With this condition, Lee et al. [91] proposed a classification scheme using Bayes maximum likelihood or minimum distance (MD) classifier. In practice, each class is characterized by an elementary covariance matrix derived from training samples, and each pixel is classified according to the Bayes likelihood with the elementary covariance matrices under a given *a priori* probability and the complex Wishart distribution. The algorithm can be generalized to classify multi-frequency polarimetric SAR data or SAR data with only polarization intensity, and can therefore be applied to a wider range of situations [87, 88].

3.2.2 Integration of optical and SAR data

The coordination of Earth Observations (EO) data for agricultural monitoring necessitates the articulation of spatially explicit EO data requirements, including where [92], when [93], how frequently [94], over which spectral range, and at what spatial resolution these data are needed [95]. Because cropping systems are often diverse and complex, and the types of crops grown and the timing of their growth vary from region to region, the best choice of sensors to be used, the optimal number of images required, and the timing of image acquisitions are usually geographically specific. Where SAR has been used in operational national-scale crop mapping programs, it has usually been integrated with optical remote sensing data. Both optical and SAR provide unique and valuable information relating to plant growth and type, primarily due to their different wavelengths. Optical imagery acquired in the near-infrared and shortwave-infrared is sensitive to canopy biochemistry such as composition and concentration of pigments, water content, biomass, and leaf internal structure, while SAR imagery is sensitive to plant structure. SAR observations are also critical for filling gaps in the optical image record brought about by the presence of clouds during key growth stages.

The integration of optical and SAR data can be as simple as combining data from different sources into raster stacks for classification, sometimes applying mathematical transformations to fuse and enhance features or reduce data dimensionality [96–99]. In some cases, SAR data are not used directly in the classification process but are first transformed into higher-level data products. This has included the derivation of phenological metrics from SAR time series (e.g., Torbick et al. [100] and the use of SAR-based texture [101].

One of the most well-known applications of SAR in national-scale crop mapping comes from Canada. Since 2010, Agriculture and Agri-Food Canada—Canada's Ministry of Agriculture—has integrated C-band SAR (RADARSAT, Sentinel-1) with optical data streams (Landsat-5, -7 and -8, SPOT, DMC, RapidEye, and Resourcesat-1) to generate its Annual Space-Based Crop Inventory for Canada [102]. **Figure 2** shows the mapping result for 2020, which covers the agricultural land and includes all crops and a few other land cover types.

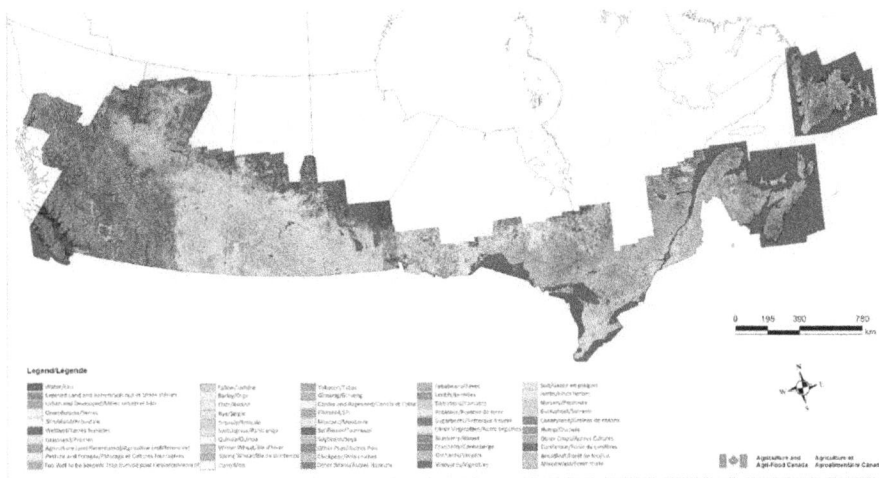

Figure 2.
National scale crop type mapping in Canada, 2020. The map is produced by Agriculture, Geomatics and Earth Observation Division, Science & Technology Branch, Agriculture, and Agri-Food Canada.

Results from research and operations suggest that optical and SAR satellite data are both required to best characterize the key crop growing (phenological) stages required for high-accuracy crop mapping at a national scale [39, 56, 80, 103–106]. The addition of dual-pol SAR has been shown to increase accuracies over the use of optical data alone by as much as 16% [42, 104]. Nonetheless, the decision to use optical and/or SAR is usually determined by the trade-off among a number of factors, including—(a) the heterogeneous and dynamic intrinsic nature of the agro-ecosystem being studied; (b) the geographical extent to be mapped; (c) the minimum mapping unit required to resolve individual fields and other meaningful ecological units (e.g., wetlands, woodlots, etc.); (d) differences in crop cycles; (e) differences in cropping practices and calendars within the same class; (f) the spectral similarity with other land cover classes; (g) the engineering constraints of the remote sensing systems (i.e., swath size; spatial, temporal, spectral and radiometric resolutions; cloud coverage for optical systems), and (h) data availability (i.e., open, fee-based).

3.3 Crop parameter estimation and growth condition monitoring

Microwave scattering, represented by both intensity and phase characteristics, changes with variations in the structure of crop canopies and canopy water content. Canopy structure and water content vary as crops develop and are thus indicative of crop development and productivity. **Figure 3** shows seasonal variation of radar backscattering intensity of annual crops over a growing season in an agricultural region in northern Ontario, Canada, using dual-polarization C-band SAR data acquired by Sentinel-1 in 2019. Both VV and VH polarizations of annual crops show obvious seasonal variation patterns characteristic to crop development cycle, whereas that of forest targets (the two dotted lines) remain at a relatively stable and higher level throughout the season. This clearly shows a positive correlation between radar backscattering intensity and crop live biomass, based on which different crop parameters can be estimated from SAR data.

The potential of SAR for supporting crop growth monitoring through the quantitative estimation of crop parameters—such as Leaf (or Plant) Area Index

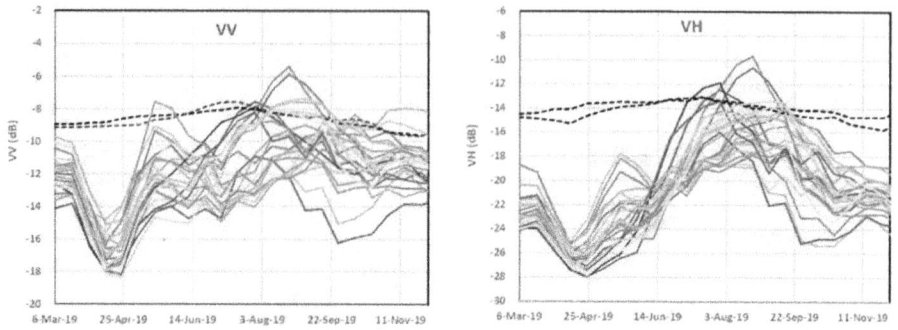

Figure 3.
Seasonal profiles of radar backscattering intensity for annual crops in northern Ontario, Canada, using C-band SAR data acquired by Sentinel-1 in 2019. The two dotted lines represent two forest patches.

(LAI or PAI), plant height and density, fresh and dry biomass, and plant water content—depends on SAR sensor characteristics (frequency, polarization, and incident angle), crop type, and growth stage [107, 108].

The characteristics of a SAR determine the depth to which a pulse of microwave energy can penetrate a plant canopy and, in turn, influence the ability to determine canopy conditions from SAR observations. Because of this, the optimal choice of SAR frequency will vary over time, depending on canopy type and growth stage, and thus the use of multiple SAR frequencies for crop mapping, where available, is recommended. SAR scattering is also polarization-dependent [44, 109, 110]. Overall, the best polarization for crop characterization has been the linear cross polarization (either HV or VH) [110, 111]. This is mainly due to re-polarization that occurs during multiple scattering within targets with complex structures, such as crop canopies consisting of randomly oriented and distributed stems and leaves [112]. Using RADARSAT-2 SAR data, Liao et al. [113] studied the sensitivity of C-band SAR polarimetric parameters for the estimation of crop height and fractional vegetation cover. They found that cross polarization or combinations of dual polarizations (HH-VV or HV-VV) were strongly correlated with crop height and fractional cover of broadleaf crops, such as corn, with degraded performance toward the later growing stages. For narrow-leaf crops, such as wheat, the sensitivity of SAR parameters to crop height and cover fraction was relatively low or inconsistent. Wali et al. [114] assessed Sentinel-1 C-band SAR VV and VH backscatter for estimating biophysical parameters of rice, including plant height, green vegetation cover, LAI, and total dry biomass. The results of this study showed that both VH and VV were strongly and linearly correlated with biophysical parameters until backscatter saturated during the mid-reproductive stage (60 days after transplanting), and the beginning of the reproductive stage for VV (though VH showed stronger correlations in most cases). Chauhan et al. [115] were able to obtain better estimates of vegetation parameters by accounting for soil backscatter effects. Other studies include those by Xie et al. [110], who demonstrated the capability of RADARSAT-2 polarimetric SAR variables for crop height estimation, and Hosseini et al. [116], who used WCM and SVM to estimate LAI using RADARSAT-2 SAR intensity collected over multiple international sites (Argentina, Canada, Germany, India, Poland, Ukraine, and the U.S).

Polarimetric SAR allows the complete scattering characteristics of crop canopies to be determined, and parameters derived from these complex data can improve estimates of crop conditions. Many recent examples of this come from studies over agricultural regions in Canada. Using C-band RADARSAT-2 polarimetric data, Wiseman et al. [117] extracted and evaluated 21 polarimetric parameters to estimate dry biomass for canola, corn, soybean, and spring wheat crops. This study found that most SAR parameters were significantly correlated with dry biomass accumulation, while several proved to be good indicators of changes in crop structure and phenology. For instance, four SAR responses (linear HV and circular LR backscatter, volume scattering, and pedestal height) increased during canola ripening. However, as canola flowered, the importance of these parameters declined. Homayouni et al. [118] used the ratio of volume-to-surface scattering derived from C-band RADARSAT-2 polarimetric data to monitor the growth of canola, corn, spring wheat, and soybeans fields in western Canada. They found that this ratio was strongly correlated with optical vegetation indices (e.g., the normalized difference vegetation index NDVI, and the Soil Adjusted Vegetation Index SAVI). Using time-series RADARSAT-2 polarimetric data, and RapidEye optical imagery, Jiao et al. [119] applied a semi-empirical Canopy Structure Dynamics Model, Growing Degree Days, and SAR parameters calibrated to optical NDVI to derive daily estimates of canola crop condition over an entire growing season. Correlations (R values) of 0.63–0.84 were reported when SAR parameters were related to optical NDVI, with results varying among three growing seasons.

A growing literature focusing on SAR-based vegetation indices demonstrates the potential of such techniques. Kim and van Zyl [120] proposed a radar vegetation index (RVI) based on the SAR backscatter intensities at VV, HH, and VH polarizations, which has since been simplified to accommodate data obtained from dual-polarized systems [121, 122]. Using Sentinel-1 observations, Periasamy [123] proposed a Dual Polarization SAR Vegetation index (DPSVI) by exploiting the data distribution of VV and VH backscatter coefficients in two-dimensional space. Such radar indices show strong potential for the better discrimination of bare soil from vegetation, as well as for crop structural parameter estimation. Other SAR-based vegetation indices include the SAR simple difference (SSD) index, applied to estimate rice yield in China, and based on the difference in Sentinel-1 VH backscatter between the end of the rice tillering stage and the end of grain filling stage [124]. Results of the study showed a strong exponential relationship between the SSDVH and rice yield.

Other applications of SAR for crop parameter estimation and growth condition monitoring include its use in radiative models. Attema and Ulaby [4] developed a Water Cloud Model (WCM) to simulate SAR backscatter from the crop-soil system as an incoherent sum of contributions from plants and background soil after a two-way attenuation by canopies. Through time, the model has been modified to reflect different approaches to the interaction and parameterization of soil and vegetation contributions. For example, various studies have used LAI, canopy water content, and biomass to characterize the vegetation component in the WCM [7, 8, 10, 116].

3.4 Change detection

In the context of this article, the objective of change detection using remote sensing data is to identify and characterize changes in agricultural land cover and/or use (e.g., conversions from one crop class to another) or changes in condition within a land cover and/or use (e.g., modifications within a crop class) over a specified period of time. These changes can be described as—(a) binary change/non-change (e.g., harvest); (b) from-to trajectories (e.g., forest to cropland conversion);

(c) causes of change (e.g., fire, flooding); and (d) continuous variable change (e.g., reduced productivity within a class due to insect infestation or drought) [125]. Understanding the types of change sought is critical for selecting suitable remote sensing data sources, determining processing methods, and developing and implementing robust and effective change detection algorithms.

For agricultural resource management, it is important to detect intra-annual landscape changes, such as changes in crop phenology [17, 20, 33], field operations [19, 21], and field conditions [126, 127]. This type of monitoring requires dense remote sensing time series that usually cannot be fulfilled using optical data alone due to the presence of the cloud. As a result, spatially and temporally comprehensive and consistent coverages from operational wide-swath SAR satellites will continue to be a critical source of free and open SAR data for national-scale change detection. As new SAR missions are launched and existing missions expand, multi-frequency SAR is expected to play an increasingly important role in monitoring and measuring change on agricultural landscapes. The application of SAR for within-season change detection will require well-calibrated data from multiple satellites within a constellation, if satellites from different constellations are used together.

Change detection using SAR backscatter—as opposed to its indirect detection from SAR-derived value-added products such as crop type maps or modeled biophysical parameters—belong to one of two broad types. These are Incoherent Change Detection (ICD) and Coherent Change Detection (CCD) [128]. ICD methods identify changes in mean backscatter intensity without considering SAR phase information. Here, the difference can be calculated as a ratio, a log ratio (LR), a mean ratio (MR), the normalized compression distance [129], or using pointwise approaches based on graph theory [130], convolutional neural networks [131], or the generalized likelihood ratio test (GLRT) [132]. In comparison, CCD methods identify change based on the complex conjugate correlation coefficient of the two images, thus taking into consideration of both backscatter intensity and phase. If dense stacks of time-series SAR images are available, changes can be inferred from these methods. For example, Shang et al. [21] used CCD to detect crop seeding and harvest using time-series Sentinel-1 SAR. The study integrated time-series coherence and VH backscatter intensity to detect changes at the beginning and at the end of a growing season, with the assumption that coherence is comparatively higher before crops emerge and after crop harvest. **Figure 4** shows the example for mapping crop seeding dates, and details of the approach are given in Shang et al. [21].

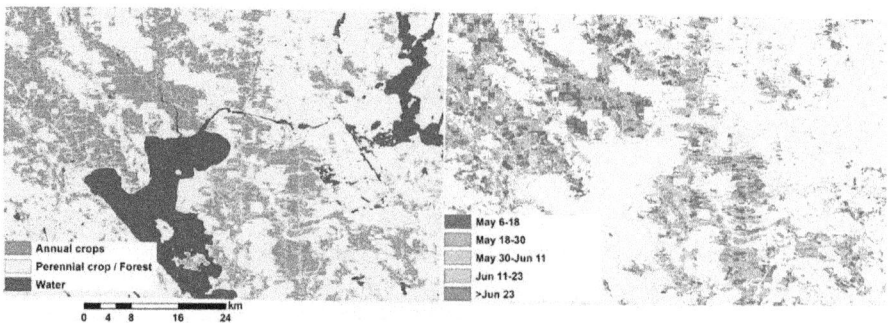

Figure 4.
Estimation of crop seeding dates through change detection using C-band SAR data acquired by Sentinel-1. Left: detection of annual crop fields using a simple threshold of seasonal variation amplitude of VH; right: mapping of crop seeding dates.

4. Summary and perspective

Timely and continuous observations from satellite systems are critical for providing the data and information required by decision-makers to manage agricultural lands. High-quality satellite observations can be obtained from SAR sensors; however, they must be collected at a spatial resolution that allows sufficient detail to be resolved, and at times, during the growing season, that coincides with the key growth stages of crops being assessed. The most accurate detailed national crop mapping generally occurs when moderate-resolution spectrally rich time series are acquired that contain no gaps.

Because of its near all-weather capacity, SAR technology has been shown to be particularly useful in agricultural monitoring, especially in regions with frequent cloud cover. The agricultural applications summarized in this article cover examples of information extraction for crops. Despite the gains made over the past 15 years in methods for crop monitoring from SAR, some challenges remain. A major challenge is the separation of backscattering signals from soils and crops, where it is difficult to differentiate the geometrical and dielectric properties of these two targets. While theoretical and semi-empirical models have been developed to simulate backscattering signals, model inversion for solving surface parameters with high accuracy remains a challenge. Much attention has been focused on the integration of SAR and optical remote sensing for improving target parameter retrieval accuracies. With temporally dense imaging capabilities of current and future satellite SAR systems, changes in agricultural land should be more accurately detected. Methods for change detection based on SAR and optical time series show large future potential.

Future opportunities for the use of SAR in agricultural monitoring will come from the adoption of new and improved satellite missions that, in combination or isolation, will allow a better characterization of crop-specific growth cycles at the field level. Of particular interest is the integration of SAR imagery acquired at multiple frequencies, especially if these multi-frequency data sets are collected in wide swaths, with consistent coverages, and under open data policies. However, this will not be without a challenge. The ability of national mapping agencies to incorporate this information in a timely and efficient manner will require significant investment in information technology infrastructure to facilitate the processing of significantly greater volumes of data.

Author details

Jiali Shang[1*], Jiangui Liu[1], Zhongxin Chen[2], Heather McNairn[1]
and Andrew Davidson[1]

1 Agriculture and Agri-Food Canada, Ottawa, Ontario, Canada

2 Food and Agriculture Organization of the United Nations, Rome, Italy

*Address all correspondence to: jiali.shang@agr.gc.ca

IntechOpen

References

[1] Fung AK, Li Z, Chen KS. Backscattering from a randomly rough dielectric surface. IEEE Transactions on Geoscience and Remote Sensing. 1992;**30**:356-369

[2] Shi J, Chen KS, Li Q, Jackson TJ, O'Neill PE, Tsang L. A parameterized surface reflectivity model and estimation of bare-surface soil moisture with L-band radiometer. IEEE Transactions on Geoscience and Remote Sensing. 2002;**40**:2674-2686

[3] Ulaby FT, Sarabandi K, McDonald K, Whitt M, Craig Dobson M. Michigan microwave canopy scattering model. International Journal of Remote Sensing. 1990;**11**:1223-1253

[4] Attema EPW, Ulaby FT. Vegetation modeled as a water cloud. Radio Science. 1978;**13**:357-364

[5] Oh Y, Sarabandi K, Ulaby FT. An empirical model and an inversion technique for radar scattering from bare soil surfaces. IEEE Transactions on Geoscience and Remote Sensing. 1992;**30**:370-381

[6] Dubois PC, van Zyl J, Engman T. Measuring soil moisture with imaging radars. IEEE Transactions on Geoscience and Remote Sensing. 1995;**33**:915-926

[7] Fieuzal R, Baup F. Estimation of leaf area index and crop height of sunflowers using multi-temporal optical and SAR satellite data. International Journal of Remote Sensing. 2016;**37**: 2780-2809

[8] Hosseini M, McNairn H. Using multi-polarization C- and L-band synthetic aperture radar to estimate biomass and soil moisture of wheat fields. International Journal of Applied Earth Observation and Geoinformation. 2017;**58**:50-64

[9] Hosseini M, McNairn H, Merzouki A, Pacheco A. Estimation of Leaf Area Index (LAI) in corn and soybeans using multi-polarization C- and L-band radar data. Remote Sensing of Environment. 2015;**170**:77-89

[10] Mandal D, Kumar V, Lopez-Sanchez JM, Bhattacharya A, McNairn H, Rao YS. Crop biophysical parameter retrieval from Sentinel-1 SAR data with a multi-target inversion of Water Cloud Model. International Journal of Remote Sensing. 2020;**41**: 5503-5524

[11] Zribi M, Baghdadi N, Holah N, Fafin O. New methodology for soil surface moisture estimation and its application to ENVISAT-ASAR multi-incidence data inversion. Remote Sensing of Environment. 2005;**96**: 485-496

[12] Zribi M, Dechambre M. A new empirical model to retrieve soil moisture and roughness from C-band radar data. Remote Sensing of Environment. 2002;**84**:42-52

[13] Cloude SR, Pottier E. A review of target decomposition theorems in radar polarimetry. IEEE Transactions on Geoscience and Remote Sensing. 1996;**34**:498-518

[14] Touzi R, Boerner WM, Lee JS, Lueneburg E. A review of polarimetry in the context of synthetic aperture radar: Concepts and information extraction. Canadian Journal of Remote Sensing. 2004;**30**:380-407

[15] Cloude SR, Pottier E. An entropy based classification scheme for land applications of polarimetric SAR. IEEE Transactions on Geoscience and Remote Sensing. 1997;**35**:68-78

[16] Macrì Pellizzeri T. Classification of polarimetric SAR images of suburban

areas using joint annealed segmentation and "H/A/α" polarimetric decomposition. ISPRS Journal of Photogrammetry and Remote Sensing. 2003;**58**:55-70

[17] Canisius F, Shang J, Liu J, Huang X, Ma B, Jiao X, et al. Tracking crop phenological development using multi-temporal polarimetric Radarsat-2 data. Remote Sensing of Environment. 2018;**210**:508-518

[18] Huang X, Ziniti B, Cosh MH, Reba M, Wang J, Torbick N. Field-scale soil moisture retrieval using palsar-2 polarimetric decomposition and machine learning. Agronomy. 2021;**11**(1):35

[19] Kavats O, Khramov D, Sergieieva K, Vasyliev V. Monitoring harvesting by time series of Sentinel-1 SAR data. Remote Sensing. 2019;**11**(21):2496

[20] Schlund M, Erasmi S. Sentinel-1 time series data for monitoring the phenology of winter wheat. Remote Sensing of Environment. 2020;**246**:111814

[21] Shang J, Liu J, Poncos V, Geng X, Qian B, Chen Q, et al. Detection of crop seeding and harvest through analysis of time-series Sentinel-1 interferometric SAR data. Remote Sensing. 2020;**12**(10):1551

[22] Bouman BAM. Crop parameter estimation from ground-based x-band (3-cm wave) radar backscattering data. Remote Sensing of Environment. 1991;**37**:193-205

[23] Brisco B, Brown RJ, Gairns JG, Snider B. Temporal ground-based scatterometer observations of crops in Western Canada. Canadian Journal of Remote Sensing. 1992;**18**:14-21

[24] Inoue Y, Kurosu T, Maeno H, Uratsuka S, Kozu T, Dabrowska-Zielinska K, et al. Season-long daily measurements of multifrequency

(Ka, Ku, X, C, and L) and full-polarization backscatter signatures over paddy rice field and their relationship with biological variables. Remote Sensing of Environment. 2002;**81**:194-204

[25] Krul L. Some results of microwave remote sensing research in the netherlands with a view to land applications in the 1990s. International Journal of Remote Sensing. 1988;**9**: 1553-1563

[26] Ulaby FT. Radar response to vegetation. IEEE Transactions on Antennas and Propagation. 1975;**23**: 36-45

[27] Ulaby FT, Wilson EA. Microwave attenuation properties of vegetation canopies. IEEE Transactions on Geoscience and Remote Sensing. 1985;**GE-23**:746-753

[28] Steele-Dunne SC, McNairn H, Monsivais-Huertero A, Judge J, Liu PW, Papathanassiou K. Radar remote sensing of agricultural canopies: A review. IEEE Journal of Selected Topics in Applied Earth Observations and Remote Sensing. 2017;**10**:2249-2273

[29] Ao D, Dumitru CO, Schwarz G, Datcu M. Dialectical GAN for SAR image translation: From sentinel-1 to TerraSAR-X. Remote Sensing. 2018;**10**(10):1597

[30] Florian C, Cignani R, Santarelli A, Filicori F. Design of 40-W AlGaN/GaN MMIC high power amplifiers for C-Band SAR applications. IEEE Transactions on Microwave Theory and Techniques. 2013;**61**:4492-4504

[31] De Bernardis CG, Vicente-Guijalba F, Martinez-Marin T, Lopez-Sanchez JM. Estimation of key dates and stages in rice crops using dual-polarization SAR time series and a particle filtering approach. IEEE Journal of Selected

Topics in Applied Earth Observations and Remote Sensing. 2015;**8**:1008-1018

[32] Dey S, Bhogapurapu N, Bhattacharya A, Mandal D, Lopez-Sanchez JM, McNairn H, et al. Rice phenology mapping using novel target characterization parameters from polarimetric SAR data. International Journal of Remote Sensing. 2021;**42**:5519-5543

[33] McNairn H, Jiao X, Pacheco A, Sinha A, Tan W, Li Y. Estimating canola phenology using synthetic aperture radar. Remote Sensing of Environment. 2018;**219**:196-205

[34] Wang H, Magagi R, Goïta K, Trudel M, McNairn H, Powers J. Crop phenology retrieval via polarimetric SAR decomposition and Random Forest algorithm. Remote Sensing of Environment. 2019;**231**:111234

[35] Bargiel D. A new method for crop classification combining time series of radar images and crop phenology information. Remote Sensing of Environment. 2017;**198**:369-383

[36] Choudhury I, Chakraborty M, Santra SC, Parihar JS. Methodology to classify rice cultural types based on water regimes using multi-temporal RADARSAT-1 data. International Journal of Remote Sensing. 2012;**33**: 4135-4160

[37] Son NT, Chen CF, Chen CR, Toscano P, Cheng YS, Guo HY, et al. A phenological object-based approach for rice crop classification using time-series Sentinel-1 synthetic aperture radar (SAR) data in Taiwan. International Journal of Remote Sensing. 2021;**42**: 2722-2739

[38] Yusoff NM, Muharam FM, Takeuchi W, Darmawan S, Abd Razak MH. Phenology and classification of abandoned agricultural land based on ALOS-1 and 2 PALSAR multi-temporal measurements. International Journal of Digital Earth. 2017;**10**:155-174

[39] Liu J, Huffman T, Shang J, Qian B, Dong T, Zhang Y. Identifying major crop types in eastern canada using a fuzzy decision tree classifier and phenological indicators derived from time series MODIS data. Canadian Journal of Remote Sensing. 2016;**42**:259-273

[40] Zhong L, Hawkins T, Biging G, Gong P. A phenology-based approach to map crop types in the San Joaquin Valley, California. International Journal of Remote Sensing. 2011;**32**:7777-7804

[41] Lin YC, Sarabandi K. A Monte Carlo coherent scattering model for forest canopies using fractal-generated trees. IEEE Transactions on Geoscience and Remote Sensing. 1999;**37**:440-451

[42] McNairn H, Shang J, Jiao X, Champagne C. The contribution of ALOS PALSAR multipolarization and polarimetric data to crop classification. IEEE Transactions on Geoscience and Remote Sensing. 2009;**47**:3981-3992

[43] Wang C, Wu J, Zhang Y, Pan G, Qi J, Salas WA. Characterizing L-band scattering of paddy rice in southeast China with radiative transfer model and multitemporal ALOS/PALSAR imagery. IEEE Transactions on Geoscience and Remote Sensing. 2009;**47**:988-998

[44] Lee JS, Pottier E. Polarimetric Radar Imaging: From Basics to Applications. 2nd ed. Taylor and Francis; 2017. pp. 475. ISBN-13: 9781466585393

[45] Touzi R, Charbonneau F. Characterization of target symmetric scattering using polarimetric SARs. IEEE Transactions on Geoscience and Remote Sensing. 2002;**40**:2507-2516

[46] Raney RK. Hybrid-polarity SAR architecture. IEEE Transactions on Geoscience and Remote Sensing. 2007;**45**:3397-3404

[47] Charbonneau FT, Brisco B, Raney RK, McNairn H, Liu C, Vachon PW, et al. Compact polarimetry overview and applications assessment. Canadian Journal of Remote Sensing. 2010;**36**:S298-S315

[48] Souyris JC, Imbo P, Fjørtoft R, Mingot S, Lee JS. Compact polarimetry based on symmetry properties of geophysical media: The π/4 mode. IEEE Transactions on Geoscience and Remote Sensing. 2005;**43**:634-645

[49] Cloude SR, Goodenough DG, Chen H. Compact decomposition theory. IEEE Geoscience and Remote Sensing Letters. 2012;**9**:28-32

[50] Ponnurangam GG, Rao YS. The application of compact polarimetric decomposition algorithms to L-band PolSAR data in agricultural areas. International Journal of Remote Sensing. 2018;**39**:8337-8360

[51] Baghdadi N, Boyer N, Todoroff P, El Hajj M, Bégué A. Potential of SAR sensors TerraSAR-X, ASAR/ENVISAT and PALSAR/ALOS for monitoring sugarcane crops on Reunion Island. Remote Sensing of Environment. 2009;**113**:1724-1738

[52] Chen KS, Huang WP, Tsay DH, Amar F. Classification of multifrequency polarimetric SAR imagery using a dynamic learning neural network. IEEE Transactions on Geoscience and Remote Sensing. 1996;**34**:814-820

[53] Dobson MC, Pierce LE, Ulaby FT. Knowledge-based land-cover classification using ERS-1/JERS-1 SAR composites. IEEE Transactions on Geoscience and Remote Sensing. 1996;**34**:83-99

[54] Ferrazzoli P, Paloscia S, Pampaloni P, Schiavon G, Sigismondi S, Solimini D. The potential of multifrequency polarimetric SAR in

assessing agricultural and arboreous biomass. IEEE Transactions on Geoscience and Remote Sensing. 1997;**35**:5-17

[55] Hoekman DH, Vissers MAM. A new polarimetric classification approach evaluated for agricultural crops. IEEE Transactions on Geoscience and Remote Sensing. 2003;**41**:2881-2889

[56] Shang J, McNairn H, Champagne C, Jiao X. Application of Multi-Frequency Synthetic Aperture Radar (SAR) in Crop Classification. In: Jedlovec G, editor. Advances in Geoscience and Remote Sensing. London: IntechOpen; 2009. DOI: 10.5772/46139

[57] Skriver H. Crop classification by multitemporal C- and L-band single- and dual-polarization and fully polarimetric SAR. IEEE Transactions on Geoscience and Remote Sensing. 2012;**50**:2138-2149

[58] Kraatz S, Torbick N, Jiao X, Huang X, Robertson LD, Davidson A, et al. Comparison between dense l-band and c-band synthetic aperture radar (SAR) time series for crop area mapping over a nisar calibration-validation site. Agronomy. 2021;**11**(2):273

[59] Karam MA, Amar F, Fung AK, Mougin E, Lopes A, Le Vine DM, et al. A microwave polarimetric scattering model for forest canopies based on vector radiative transfer theory. Remote Sensing of Environment. 1995;**53**:16-30

[60] Touré A, Thomson KPB, Edwards G. Adaptation of the MIMICS backscattering model to the agricultural context—wheat and canola at L and C bands. IEEE Transactions on Geoscience and Remote Sensing. 1994;**32**:47-61

[61] Prevot L, Dechambre M, Taconet O, Vidal-Madjar D, Normand M, Galle S. Estimating the

characteristics of vegetation canopies with airborne radar measurements. International Journal of Remote Sensing. 1993;**14**:2803-2818

[62] Poirier S, Thomson KP, Condal A, Brown RJ. SAR applications in agriculture: A comparison of steep and shallow mode (30° and 53° incidence angles) data. International Journal of Remote Sensing. 1989;**10**:1085-1092

[63] Kothapalli Venkata R, Poloju S, Mullapudi Venkata Rama SS, Gogineni A, Prabir Kumar D, Allakki Venkata R, et al. Multi-incidence angle RISAT-1 hybrid polarimetric SAR data for large area mapping of maize crop—a case study in Khagaria district, Bihar, India. International Journal of Remote Sensing. 2017;**38**:5487-5501

[64] Xu S, Qi Z, Li X, Yeh AGO. Investigation of the effect of the incidence angle on land cover classification using fully polarimetric SAR images. International Journal of Remote Sensing. 2019;**40**:1576-1593

[65] Kumar L, Sinha P, Brown JF, Ramsey RD, Rigge M, Stam CA, et al. Characterization, mapping, and monitoring of rangelands: Methods and approaches. In: Land Resources Monitoring, Modeling, and Mapping with Remote Sensing. 1st ed. Boca Raton: CRC Press; 2015. p. 885. DOI: 10.1201/b19322

[66] Lu D, Weng Q. A survey of image classification methods and techniques for improving classification performance. International Journal of Remote Sensing. 2007;**28**:823-870

[67] Richards JA, Jia X. Remote Sensing Digital Image Analysis: An Introduction. 4th ed. New York: Springer; 2006. 439p. ISBN: 3540251286

[68] Abdulaziz AM, Hurtado JM, Al-Douri R. Application of multitemporal Landsat data to monitor land cover

changes in the Eastern Nile Delta region, Egypt. International Journal of Remote Sensing. 2009;**30**:2977-2996

[69] Kamusoko C, Aniya M. Hybrid classification of Landsat data and GIS for land use/cover change analysis of the Bindura district, Zimbabwe. International Journal of Remote Sensing. 2008;**30**:97-115

[70] Rogan J, Franklin J, Stow D, Miller J, Woodcock C, Roberts D. Mapping land-cover modifications over large areas: A comparison of machine learning algorithms. Remote Sensing of Environment. 2008;**112**:2272-2283

[71] Xiuwan C. Using remote sensing and GIS to analyse land cover change and its impacts on regional sustainable development. International Journal of Remote Sensing. 2002;**23**:107-124

[72] Atkinson PM, Tatnall ARL. Introduction neural networks in remote sensing. International Journal of Remote Sensing. 1997;**18**:699-709

[73] Mas JF, Flores JJ. The application of artificial neural networks to the analysis of remotely sensed data. International Journal of Remote Sensing. 2007;**29**: 617-663

[74] Rigol-Sanchez JP, Chica-Olmo M, Abarca-Hernandez F. Artificial neural networks as a tool for mineral potential mapping with GIS. International Journal of Remote Sensing. 2003;**24**:1151-1156

[75] Rumelhart DE, Hinton GE, Williams RJ. Learning representations by back-propagating errors. Nature. 1986;**323**:533-536

[76] Cortes C, Vapnik V. Support-vector networks. Machine Learning. 1995;**20**:273-297

[77] Kavzoglu T, Colkesen I. A kernel functions analysis for support vector machines for land cover classification.

International Journal of Applied Earth Observation and Geoinformation. 2009;**11**:352-359

[78] Maxwell AE, Warner TA, Fang F. Implementation of machine-learning classification in remote sensing: An applied review. International Journal of Remote Sensing. 2018;**39**:2784-2817

[79] Zuo R, Carranza EJM. Support vector machine: A tool for mapping mineral prospectivity. Computers and Geosciences. 2011;**37**:1967-1975

[80] Champagne C, McNairn H, Daneshfar B, Shang J. A bootstrap method for assessing classification accuracy and confidence for agricultural land use mapping in Canada. International Journal of Applied Earth Observation and Geoinformation. 2014;**29**:44-52

[81] Ghimire B, Rogan J, Galiano V, Panday P, Neeti N. An evaluation of bagging, boosting, and random forests for land-cover classification in Cape Cod, Massachusetts, USA. GIScience and Remote Sensing. 2012;**49**:623-643

[82] Khosravi I, Safari A, Homayouni S, McNairn H. Enhanced decision tree ensembles for land-cover mapping from fully polarimetric SAR data. International Journal of Remote Sensing. 2017;**38**:7138-7160

[83] Mahdianpari M, Mohammadimanesh F, McNairn H, Davidson A, Rezaee M, Salehi B, et al. Mid-season crop classification using dual-, compact-, and full-polarization in preparation for the Radarsat Constellation Mission (RCM). Remote Sensing. 2019;**11**(13):1582

[84] Zhang H, Li Q, Liu J, Du X, Dong T, McNairn H, et al. Object-based crop classification using multi-temporal SPOT-5 imagery and textural features with a Random Forest classifier. Geocarto International. 2018;**33**:1017-1035

[85] Khatami R, Mountrakis G, Stehman SV. A meta-analysis of remote sensing research on supervised pixel-based land-cover image classification processes: General guidelines for practitioners and future research. Remote Sensing of Environment. 2016;**177**:89-100

[86] Adams JR, Rowlandson TL, McKeown SJ, Berg AA, McNairn H, Sweeney SJ. Evaluating the Cloude-Pottier and Freeman-Durden scattering decompositions for distinguishing between unharvested and post-harvest agricultural fields. Canadian Journal of Remote Sensing. 2013;**39**:318-327

[87] Ferro-Famil LP, Pottier E, Lee JS. Unsupervised classification of multifrequency and fully polarimetric SAR images based on the H/A/alpha-Wishart classifier. IEEE Transactions on Geoscience and Remote Sensing. 2001;**39**:2332-2342

[88] Lee JS, Grunes MR, Ainsworth TL, Du LJ, Schuler DL, Cloude SR. Unsupervised classification using polarimetric decomposition and the complex Wishart classifier. IEEE Transactions on Geoscience and Remote Sensing. 1999;**37**:2249-2258

[89] Park SE, Moon WM. Unsupervised classification of scattering mechanisms in polarimetrie SAR data using fuzzy logic in entropy and alpha plane. IEEE Transactions on Geoscience and Remote Sensing. 2007;**45**:2652-2664

[90] Tan CP, Ewe HT, Chuah HT. Agricultural crop-type classification of multi-polarization SAR images using a hybrid entropy decomposition and support vector machine technique. International Journal of Remote Sensing. 2011;**32**:7057-7071

[91] Lee JS, Grunes MR, Kwok R. Classification of multi-look polarimetric SAR imagery based on complex Wishart distribution.

International Journal of Remote Sensing. 1994;**15**:2299-2311

[92] Fritz S, See L, McCallum I, You L, Bun A, Moltchanova E, et al. Mapping global cropland and field size. Global Change Biology. 2015;**21**:1980-1992

[93] Whitcraft AK, Becker-Reshef I, Justice CO. Agricultural growing season calendars derived from MODIS surface reflectance. International Journal of Digital Earth. 2015;**8**:173-197

[94] Whitcraft AK, Vermote EF, Becker-Reshef I, Justice CO. Cloud cover throughout the agricultural growing season: Impacts on passive optical earth observations. Remote Sensing of Environment. 2015;**156**:438-447

[95] Whitcraft AK, Becker-Reshef I, Justice CO. A framework for defining spatially explicit earth observation requirements for a global agricultural monitoring initiative (GEOGLAM). Remote Sensing. 2015;**7**:1461-1481

[96] Abdikan S, Bilgin G, Sanli FB, Uslu E, Ustuner M. Enhancing land use classification with fusing dual-polarized TerraSAR-X and multispectral RapidEye data. Journal of Applied Remote Sensing. 2015;**9**(1):15125

[97] Gibril MBA, Bakar SA, Yao K, Idrees MO, Pradhan B. Fusion of RADARSAT-2 and multispectral optical remote sensing data for LULC extraction in a tropical agricultural area. Geocarto International. 2017;**32**:735-748

[98] Hong G, Zhang A, Zhou F, Brisco B. Integration of optical and synthetic aperture radar (SAR) images to differentiate grassland and alfalfa in Prairie area. International Journal of Applied Earth Observation and Geoinformation. 2014;**28**:12-19

[99] Zhang J. Multi-source remote sensing data fusion: Status and trends.

International Journal of Image and Data Fusion. 2010;**1**:5-24

[100] Torbick N, Chowdhury D, Salas W, Qi J. Monitoring rice agriculture across myanmar using time series Sentinel-1 assisted by Landsat-8 and PALSAR-2. Remote Sensing. 2017;**9**(2):119

[101] Inglada J, Vincent A, Arias M, Marais-Sicre C. Improved early crop type identification by joint use of high temporal resolution SAR and optical image time series. Remote Sensing. 2016;**8**(5):362

[102] Davidson AM, Fisette T, McNairn H, Daneshfar B. Detailed crop mapping using remote sensing data (crop data layers). In: Delince J, editor. Handbook on Remote Sensing for Agricultural Statistics (Chapter 4). Handbook of the Global Strategy to improve Agricultural and Rural Statistics (GSARS). Rome: GSARS Handbook; 2017

[103] Deschamps B, McNairn H, Shang J, Jiao X. Towards operational radar-only crop type classification: Comparison of a traditional decision tree with a random forest classifier. Canadian Journal of Remote Sensing. 2012;**38**:60-68

[104] Fisette T, Davidson A, Daneshfar B, Rollin P, Aly Z, Campbell L. Annual Space-Based Crop Inventory for Canada: 2009-2014. Quebec City, Canada: Joint 2014 IEEE International Geoscience and Remote Sensing Symposium (IGARSS 2014) and the 35th Canadian Symposium on Remote Sensing (CSRS 2014), Quebec Convention Centre; 2014. pp. 5095-5098

[105] Jiao X, Kovacs JM, Shang J, McNairn H, Walters D, Ma B, et al. Object-oriented crop mapping and monitoring using multi-temporal polarimetric RADARSAT-2 data. ISPRS Journal of Photogrammetry and Remote Sensing. 2014;**96**:38-46

[106] McNairn H, Champagne C, Shang J, Holmstrom D, Reichert G. Integration of

optical and synthetic aperture radar (SAR) imagery for delivering operational annual crop inventories. ISPRS Journal of Photogrammetry and Remote Sensing. 2009;**64**:434-449

[107] McNairn H, Brisco B. The application of C-band polarimetric SAR for agriculture: A review. Canadian Journal of Remote Sensing. 2004;**30**: 525-542

[108] Wigneron JP, Ferrazzoli P, Olioso A, Bertuzzi P, Chanzy A. A simple approach to monitor crop biomass from C-band radar data. Remote Sensing of Environment. 1999;**69**:179-188

[109] Cloude S. Polarisation: Applications in Remote Sensing. Oxford Scholarship Online; 2009. DOI: 10.1093/acprof:oso/9780199569731. 001.0001. ISBN-13: 9780199569731

[110] Xie Q, Wang J, Lopez-Sanchez JM, Peng X, Liao C, Shang J, et al. Crop height estimation of corn from multi-year radarsat-2 polarimetric observables using machine learning. Remote Sensing. 2021;**13**:1-19

[111] Karjalainen M, Kaartinen H, Hyyppä J. Agricultural monitoring using envisat alternating polarization SAR images. Photogrammetric Engineering and Remote Sensing. 2008;**74**:117-126

[112] Haldar D, Verma A, Pal O. Biophysical parameters retrieval and sensitivity analysis of rabi crops (mustard and wheat) from structural perspective. Progress in Electromagnetics Research C. 2020;**106**:61-75

[113] Liao C, Wang J, Shang J, Huang X, Liu J, Huffman T. Sensitivity study of radarsat-2 polarimetric SAR to crop height and fractional vegetation cover of corn and wheat. International Journal of Remote Sensing. 2018;**39**:1475-1490

[114] Wali E, Tasumi M, Moriyama M. Combination of linear regression lines to understand the response of sentinel-1 dual polarization SAR data with crop phenology-case study in Miyazaki, Japan. Remote Sensing. 2020;**12**(1):189

[115] Chauhan S, Srivastava HS, Patel P. Wheat crop biophysical parameters retrieval using hybrid-polarized RISAT-1 SAR data. Remote Sensing of Environment. 2018;**216**:28-43

[116] Hosseini M, McNairn H, Mitchell S, Robertson LD, Davidson A, Ahmadian N, et al. A comparison between support vector machine and water cloud model for estimating crop leaf area index. Remote Sensing. 2021;**13**(7):1348

[117] Wiseman G, McNairn H, Homayouni S, Shang J. RADARSAT-2 Polarimetric SAR response to crop biomass for agricultural production monitoring. IEEE Journal of Selected Topics in Applied Earth Observations and Remote Sensing. 2014;**7**:4461-4471

[118] Homayouni S, McNairn H, Hosseini M, Jiao X, Powers J. Quad and compact multitemporal C-band PolSAR observations for crop characterization and monitoring. International Journal of Applied Earth Observation and Geoinformation. 2019;**74**:78-87

[119] Jiao X, McNairn H, Dingle Robertson L. Monitoring crop growth using a canopy structure dynamic model and time series of synthetic aperture radar (SAR) data. International Journal of Remote Sensing. 2021;**42**:6437-6464

[120] Kim Y, van Zyl JJ. A time-series approach to estimate soil moisture using polarimetric radar data. IEEE Transactions on Geoscience and Remote Sensing. 2009;**47**:2519-2527

[121] Mandal D, Kumar V, Ratha D, Dey S, Bhattacharya A,

Lopez-Sanchez JM, et al. Dual polarimetric radar vegetation index for crop growth monitoring using Sentinel-1 SAR data. Remote Sensing of Environment. 2020;**247**:111954

[122] Trudel M, Charbonneau F, Leconte R. Using RADARSAT-2 polarimetric and ENVISAT-ASAR dual-polarization data for estimating soil moisture over agricultural fields. Canadian Journal of Remote Sensing. 2012;**38**:514-527

[123] Periasamy S. Significance of dual polarimetric synthetic aperture radar in biomass retrieval: An attempt on Sentinel-1. Remote Sensing of Environment. 2018;**217**:537-549

[124] Wang J, Dai Q, Shang J, Jin X, Sun Q, Zhou G, et al. Field-scale rice yield estimation using sentinel-1A synthetic aperture radar (SAR) data in coastal saline region of Jiangsu Province, China. Remote Sensing. 2019;**11**(19):2274

[125] Lu D, Li G, Moran E. Current situation and needs of change detection techniques. International Journal of Image and Data Fusion. 2014;**5**:13-38

[126] Ajadi OA, Liao H, Jaacks J, Santos AD, Kumpatla SP, Patel R, et al. Landscape-scale crop lodging assessment across iowa and illinois using synthetic aperture radar (SAR) images. Remote Sensing. 2020;**12**:1-15

[127] Chauhan S, Darvishzadeh R, Boschetti M, Nelson A. Estimation of crop angle of inclination for lodged wheat using multi-sensor SAR data. Remote Sensing of Environment. 2020;**236**:111488

[128] Jung J, Yun S-H. Evaluation of coherent and incoherent landslide detection methods based on synthetic aperture radar for rapid response: A case study for the 2018 Hokkaido

Landslides. Remote Sensing. 2020;**12**(2):265

[129] Coca M, Anghel A, Datcu M. Unbiased seamless SAR image change detection based on normalized compression distance. IEEE Journal of Selected Topics in Applied Earth Observations and Remote Sensing. 2019;**12**:2088-2096

[130] Pham MT, Mercier G, Michel J. Change detection between SAR images using a pointwise approach and graph theory. IEEE Transactions on Geoscience and Remote Sensing. 2016;**54**:2020-2032

[131] Li Y, Peng C, Chen Y, Jiao L, Zhou L, Shang R. A deep learning method for change detection in synthetic aperture radar images. IEEE Transactions on Geoscience and Remote Sensing. 2019;**57**:5751-5763

[132] Zhuang H, Tan Z, Deng K, Yao G. Adaptive generalized likelihood ratio test for change detection in SAR images. IEEE Geoscience and Remote Sensing Letters. 2020;**17**:416-420

Chapter 3

Utilization of Remote Sensing Technology for Carbon Offset Identification in Malaysian Forests

Hamdan Omar, Thirupathi Rao Narayanamoorthy,
Norsheilla Mohd Johan Chuah, Nur Atikah Abu Bakar
and Muhamad Afizzul Misman

Abstract

Rapid growth of Malaysia's economy recently is often associated with various environmental disturbances, which have been contributing to depletion of forest resources and thus climate change. The need for more spaces for numerous land developments has made the existing forests suffer from deforestation. This chapter presents an overview and demonstrates how remote sensing data is used to map and quantify changes of tropical forests in Malaysia. The analysis dealt with image processing that produce seamless mosaics of optical satellite data over Malaysia, within 15 years period, with 5-year intervals. The challenges were about the production of cloud-free images over a tropical country that always covered by clouds. These datasets were used to identify eligible areas for carbon offset in land use, land use change and forestry (LULUCF) sector in Malaysia. Altogether 580 scenes of Landsat imagery were processed to complete the observation period and came out with a seamless, wall to wall images over Malaysia from year 2005 to 2020. Forests have been identified from the image classification and then classified into three major types, which are dry-inland forest, peat swamp and mangroves. Post-classification change detection technique was used to determine areas that have been undergoing conversions from forests to other land uses. Forest areas were found to have declined from about 19.3 Mil. ha (in 2005) to 18.2 Mil. ha in year 2020. Causes of deforestation have been identified and the amount of carbon dioxide (CO_2) that has been emitted due to the deforestation activity has been determined in this study. The total deforested area between years 2005 and 2020 was at 1,087,030 ha with rate of deforestation of about 72,469 ha yr.$^{-1}$ (or 0.37% yr.$^{-1}$). This has contributed to the total CO_2 emission of 689.26 Mil. Mg CO_2, with an annual rate of 45.95 Mil. Mg CO_2 yr.$^{-1}$. The study found that the use of a series satellite images from optical sensors are the most appropriate sensors to be used for monitoring of deforestation over the Malaysia region, although cloud covers are the major issue for optical imagery datasets.

Keywords: Landsat images, tropical forests, deforestation, carbon offset, climate change

1. Introduction

Tropical forests are crucial for mitigating climate change, but many forests continue to be driven from carbon sinks to sources through human activities. To support more sustainable forest uses, therefore forests carbon needs to be measured and monitored at high spatial and temporal resolution. Tropical forest is one of the key ecosystems in addressing issues relating to climate change as it known to store large amount of carbon [1]. Retrieving tropical forest carbon over large areas has been challenging since decades due to the limited data resource, accessibility, and numerous technical issues. Remote sensing has been used actively for forest carbon estimation since the last three decades and it is proven to be effective [2, 3]. Although there are issues and arguments raised in the estimation accuracy, research is continuously being carried out. Optical or synthetic aperture radar (SAR) system has its own potential in retrieving biomass, but several issues remain unaddressed. While optical remote sensing is usually hindered by cloud, SAR systems are always limited by signal saturation at high biomass levels [4]. However, optical sensors offer better solutions for biomass assessment. Various spectral signatures and several vegetation indices can be derived from multispectral images can make the interpretation of biophysical properties of forests can be carried out conveniently. These are the most significant difference between optical and SAR systems that has made optical satellite data preferable in vegetation studies.

While the world has growing demand energy sector, it is crucial that nations put a collective effort to reduce anthropogenic greenhouse (GHG) emission and limit the global warming below 2°C above pre-industrial times, thus prevent catastrophic effects of global climate change. Mitigating the consequences of global climate change may be a critical societal objective now and within the forthcoming decades. Tropical countries contribute to carbon emissions mainly through deforestation and forest degradation, which accounts for approximately 10% of the world's annual total carbon emissions [5]. National and international initiatives such as reducing emission from deforestation and forest degradation, and forest conservation (REDD+) and carbon offset are dedicated to mitigating the impacts of global warming. To achieve this objective, each nation's carbon emissions resulting from deforestation and forest degradation need to be quantified and tracked over time. At such large geographic scales, a precise, cost-effective, and high-resolution means to monitor changes in aboveground carbon stocks is needed. This chapter is focusing on the roles of space borne remote sensing, especially free-access satellite data in assessing biomass of forest in various ecosystems in Malaysia, i.e., inland dipterocarp forests, mangrove forest, and peat swamp forest.

1.1 Forests in Malaysia

Major forest types in Malaysia are lowland dipterocarp forest, hill dipterocarp forest, upper hill dipterocarp forest, oak-laurel forest, montane ericaceous forest, peat swamp forest and mangrove forest. In addition, there also smaller areas of freshwater swamp forest, melaleuca forest, heath forest, forest on limestone and forest on quartz ridges. Considering the composition of these forests in Malaysia, the types can be generalized into three types, which are inland, peat swamp and mangroves.

The forests in Malaysia are mostly dominated by trees from the Dipterocarpaceae family, hence the term 'dipterocarp' forests. The dipterocarp forest occurs on dry land just above sea level to an altitude of about 900 meters. The term also refers to the fact that most of the largest trees in this forest belong to Dipterocarpaceae family. This type of forest can be classified according to altitude into lowland dipterocarp forest, up to 300 m above sea level, and hill dipterocarp forest found in elevation of between 300 m and 750 m above sea level, and the upper dipterocarp forests, from 750 m to 1,200 m above sea level. However, in Sarawak and Sabah both the lowland and hill dipterocarp forests are known as mixed-dipterocarp forest.

Currently, lowland dipterocarp forest is very few left outside of protected areas such as parks and wildlife reserves. While most of the country was covered with lowland forest in the past, today the majority has been cleared for other land uses. The few remaining pockets are under the gazetted land as Forest Reserves. Moreover, forest in this regime is also being central attraction for timber extractions. There is a real need to put more effort in saving and protecting this precious habitat type. Fortunately, some State (i.e., Provincial) Governments have halted land clearing for agriculture. It is vital that all remaining forest areas are protected. In this way, this valuable natural habitat can be managed on a sustainable basis.

1.2 Reducing emission from deforestation and forest degradation, and forest conservation (REDD+) in Malaysia

The REDD+ mechanism was agreed at the 15th Session of the Conference of Parties (COP 15) United Nations Framework Convention on Climate Change (UNFCCC), 2009 in Copenhagen. The REDD+ mechanism includes reducing emissions from deforestation, forest degradation, conservation, sustainable management of forest and carbon stock enhancement. It was also agreed that parties implementing REDD+ would need an effective national strategy or action plan and a transparent national forest monitoring and governance system. Ultimately, this mechanism was created to provide an incentive for developing countries to protect, better manage, and wisely use their forest resources, thereby contributing to the global fight against climate change.

Following COP 15, the progress in the REDD+ negotiations have been relatively rapid, with the most significant developments occurring in the last couple of years. Seven important decisions were adopted in 2014 for REDD+ governing methodological issues on safeguards, measurement, reporting and verification (MRV), development of national forest monitoring systems, addressing drivers of deforestation, and technical assessment of reference levels. In addition, the modalities for institutional arrangements at the national level for REDD+ implementation and results-based payments were also agreed.

Consensus on REDD+ was reached at the UNFCCC's COP 15, which agreed on the need to provide positive incentives. This is followed by the Warsaw Framework for REDD+ providing guidance on all the requirements to obtain Results Based Payments (RBP). The agreed REDD+ to capture activities are (i) reduction of emissions from deforestation, (ii) reduction of emissions from forest degradation, (iii) conservation of forest carbon stocks, (iv) pursuance of sustainable management of forests, and (v) enhancement of forest carbon stock.

Malaysia's forests can be categorized according to the degree of protection and land use classification. Management of forest land falls under three broad categories, which are: (i) Protected Areas/Totally Protected Area which consist of, national and state parks, wildlife sanctuaries, and nature reserves, (ii) Permanent Reserved Forests (PRFs) /Permanent Forest Estate (PFEs)/Permanent Forest Reserves (PFR), which are primarily natural forests to be maintained and managed sustainably for production and protection, and (iii) Stateland forest which are forest land reserved for future development purposes.

REDD+ is more than just a means of assigning monetary value to forest carbon stocks. It is also about ensuring the livelihoods of those whose culture, survival and heritage depend on the forests themselves.

1.3 The Paris agreement

The Paris Agreement builds on the Convention by bringing all nations together for the first time to commit to ambitious efforts to prevent climate change and adapt to its effects, with increased support for developing countries. As a result, it sets a new direction for the global climate effort.

The main goal of Paris Agreement is to enhance the global response to the issue of climate change by keeping a global temperature rise this century not more than 2°C above pre-industrial levels and to pursue efforts to limit the temperature increase even further to 1.5°C [6]. In addition, the agreement intends to improve countries' ability to deal with the effects of climate change. Appropriate financial flows, a new technology framework, and expanded capacity building frameworks will be put in place to achieve these lofty goals, allowing developing countries and the most vulnerable countries to pursue their own national ambitions. Through a more rigorous transparency structure, the Agreement also provides for increased action and support transparency.

1.4 Nationally determined contribution (NCD)

Nationally determined contributions (NDCs) are at the core of the Paris Agreement and the achievement of these long-term goals. Malaysia intends to reduce the emission intensity of the greenhouse gas by 45% by 2030; in which corresponding to GDP [7]. In this circumstance, developed countries should involve 35% on an unconditional basis and a further 10% is condition upon receipt of climate finance for advanced technology transformation in construction capacity enhancement. In fact, this can assist in monitoring GHG.

Since the Paris Climate Agreement was signed in late 2016, governments all over the world have been submitting plans for reducing CO_2 emissions through their NDCs. The NDCs was previously known as "Intended Nationally Determined Contributions" (INDC) and it is submitted to the United Nations Framework Convention on Climate Change (UNFCCC) once the countries ratified to the Paris Agreement. Currently, 197 parties to the Convention had submitted their INDCs and 150 had ratified it including Malaysia.

There has been no baseline prediction or quantified analysis of baseline measures provided, but Malaysia's NDC indicated a 2005 as base-year emission level of 288 Mil Mg CO_2e, which includes emissions of 25 Mil Mg CO_2 from the LULUCF sector.

In 2014, Malaysia produced an Emissions Intensity Reduction Roadmap. According to the report, the country has chances across many sectors to fulfill the reduction target of a 40% decrease in GDP emissions intensity [8]. However, even if these opportunities exist, significant work would be necessary to achieve these emissions reductions, given the challenges of a 4.8 percent yearly rate for the per capita emissions between years 2000 and 2030 [9]. Energy for transportation is expected to rise at a pace of 5.3 percent per year over the next 25 years, making it the fastest-growing sector. Malaysia's ultimate energy needs are predicted to treble by 2030, compared to present levels of consumption.

While the "ambitious scenario" indicates that Malaysia would be able to meet its Paris agreement NDC reduction target, substantial assistance from international funders is required. When both LULUCF emissions and removals are included, the

GHG emission intensity per GDP in 2030 increases when compared to 2005 levels. This is since increase in removals by the LULUCF sector is much lower than the increase in emissions from the other sectors.

To achieve this target, a carbon offset project must be developed. Forest conservation is one of the options that can be explored since the forests are able to sequester CO_2 at considerable amount. Several options have been recognized by the Verified Carbon Standards (VCS) that there are 7 types of projects related to forest conservation that can be intervened as carbon offset project (**Table 1**) [10].

1.5 Carbon offset initiative

Carbon offsetting is the process of compensating for CO_2 pollution (carbon footprint) by avoiding similar pollution from occurring elsewhere. One carbon offset entail compensating for the emission of 1 Mg of CO_2 into the atmosphere by preventing the emission of 1 Mg of CO_2 somewhere on Earth. The underlying concept is that developed countries pay poor countries (or assist them in other ways) to reduce global emissions on their behalf. In theory, carbon offsetting can assist the world to combat global warming if offsets are used to fund good, long-term environmental projects that would not have occurred otherwise.

There are dozens of different techniques to reduce carbon dioxide emissions, ranging from energy efficiency and renewable energy to forest planting. The most popular projects are those involving renewable energy; the most contentious are those involving forestry [11].

Malaysia has a lengthy history of forest management. However, some forest areas have been damaged as a result of prior management practices. The cost of

No.	Project type	Definitions
1	Avoided planned deforestation	Avoided or stopping any logging/ plantation concession that involves deforestation. Carbon stock refers to the carbon stored in trees, whereas abatement refers to the net reduction in greenhouse gas emissions as a result of a project.
2	Wetland restoration and conservation	Increasing GHG removals by restoring wetland ecosystem by rewetting or avoiding the degradation of wetlands.
3	From low to high productivity forest	Convert low-productivity forests to high-productivity forests to increase carbon sequestration. Improved stocking density in low-productivity forests can help to boost carbon stores.
4	Conversion of logged forests to protected forests	Converting logged forests by eliminating harvesting of timber, biomass carbon stocks are protected, and can increase as the forest grows and/or continues to grow.
5	Reduced impact logging	Switching from conventional logging to RIL during timber harvesting. Carbon stocks can be increases by reducing damage to other trees, improve selection of trees, improve logging plan, etc.
6	Afforestation / Reforestation	Increase carbon sequestration via planting or human-assisted natural vegetation to develop, increase, or restore vegetative cover (forest or non-forest).
7	Extending the rotation age of evenly aged managed forests	Extending the forest rotation age or cutting cycle and increase carbon stocks. No fixed period of years to be extended, but generally the longer the period, the more average carbon stock increases.

Table 1.
Types of potential carbon offset project in Malaysia.

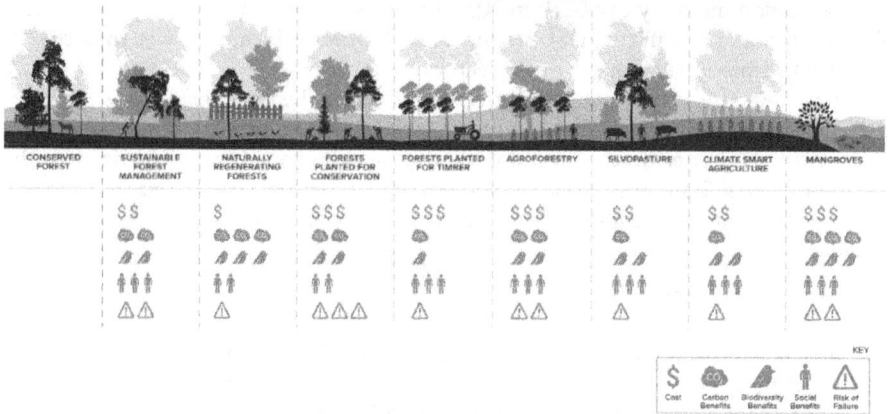

Figure 1.
Different restoration strategies can be used for different purposes and have different trade-offs.

restoring and rehabilitating these forests is high, as is the cost of caring for them. Each project type will have different implications on the cost, carbon benefits, biodiversity benefits, social benefits, and risk of failures. **Figure 1** illustrates how these implications could occur when a project type is chosen as carbon offset project [12].

2. Methodological framework

This study includes estimation of the national greenhouse gas emission trends from 2005 through 2020. This is to ensure that the LULUCF sector is the right area to venture for the carbon offset project. This is because the LULUCF sector also does emit CO_2 in various manners. Therefore, instead of CO_2 sink, it can be a source of CO_2 emission as well at some extend. To ensure that the GHGs reported in this study is comparable to UNFCCC, the estimates presented here were calculated by using methodologies consistent with those recommended in the 2006 Intergovernmental Panel on Climate Change (IPCC) Guidelines for National Greenhouse Gas Inventories. This study will not be used as regard to the carbon offset under the Paris Agreement and other international obligations but can be considered as an index that shows the extent of Malaysia mitigation measures implemented to combat the global warming.

2.1 Activity data

Remotely sensed data was used in this study for the years between 2005 and 2020 to estimate greenhouse gas emissions/removals, with 5-year interval. The reason for using these data due to its availability as independent data, it has a consistent time series and compatible with alternative data sources. In addition, the activity data obtained from Landsat satellite images are not found in any publication. Since the optical images over Malaysia are always hindered by cloud covers, a considerable number of datasets were required to produce a seamless mosaic of the images (without clouds). The Landsat images, covering the entire Malaysia that were used in this study are summarized in **Tables 2** and **3** and **Figure 2**.

2.2 Production of seamless mosaic, time series Landsat images

According to **Table 2**, only 29 scenes of Landsat images are required to produce a mosaic that cover the entire Malaysia. However, being in the tropical regions, Malaysia is always covered by clouds that is almost impossible to be removed completely. Therefore, several images acquired at different dates over the same scenes are required to produce a cloudless image. The study has set a limit of five best images of the same scenes acquired circa three years of the targeted year to be used for further processes. These images must have <30% cloud cover and acquired within the specific periods (**Table 3**). Even though Landsat has 16-day repeat cycle, which are producing about 22 images over the same scene in a year, it is still difficult to find the best five images within 3 years. This is due to the heavy cloud covers in the atmosphere of Malaysia, especially at the mountainous areas and during the monsoon season (October – February). Cloud covers in most of the scenes are ranging from 10 to 90% and therefore, the chance to obtain <30% cloud cover is very small.

However, this issue has been solved by having several good quality scenes. The clouds on these images were detected and masked by using F_Mask algorithm [13]. **Figure 3** shows the example of cloud masking a process was carried out to produce a cloudless mosaic of the scene 126/058 (**Figure 4**) that were acquired from various dates. This process is repeated for the other scenes and throughout the intervals (2005, 2010, 2015 and 2020). Altogether 580 scenes were processed to produce a seamless mosaic image for each time series. The final product is shown in **Figure 5**. Although the image looks clean, there is about ~1% of hollow pixels still appear on

Satellite	Sensor	Date of acquisition	Time series (year)
Landsat-5	Thematic Mapper (TM)	January 2004–December 2006	2005
		January 2009–December 2011	2010
Landsat-8	Operational Land Imager (OLI)	January 2014–December 2016	2015
		January 2018 – March 2020	2020

Table 2.
Satellite images that were used as activity data.

		Landsat Scene		
Region		West Malaysia	East Malaysia	
State		Peninsular Malaysia	Sarawak	Sabah
Scene ID (Path/Row)		128/055–057	121/058–059	118/055–057
		127/056–058	120/058–059	117/055–057
		126/056–059	119/057–059	116/056–057
		125/058–059	118/058–059	
Scenes required to cover Malaysia		12	9	8
Scenes acquired to produce cloud-free data		60	45	40
Scenes acquired to produce time series data (4 series)		240	180	160
Total scenes acquired	580			

Table 3.
Summary of Landsat scenes datasets required to produce seamless mosaics over the entire Malaysia.

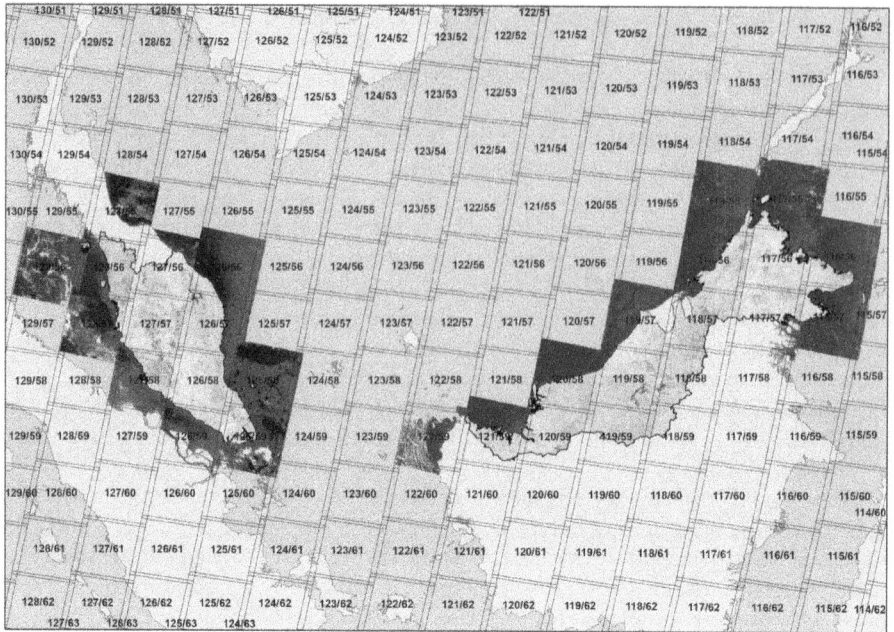

Figure 2.
Landsat scenes coverages over the entire Malaysia.

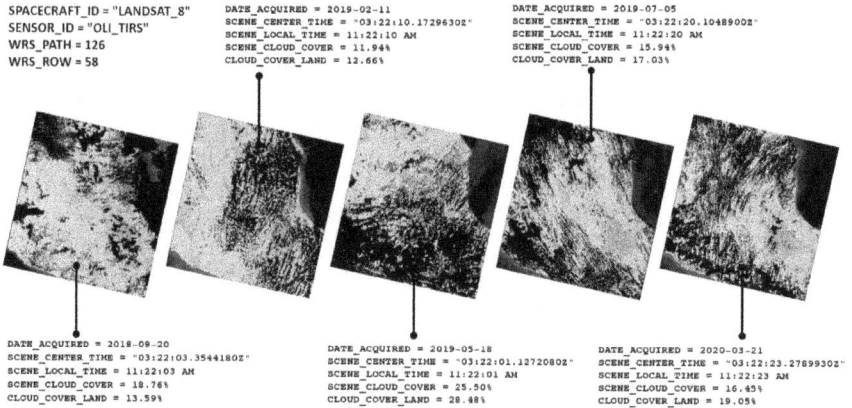

Figure 3.
Cloud masking process of Landsat scene 126/058.

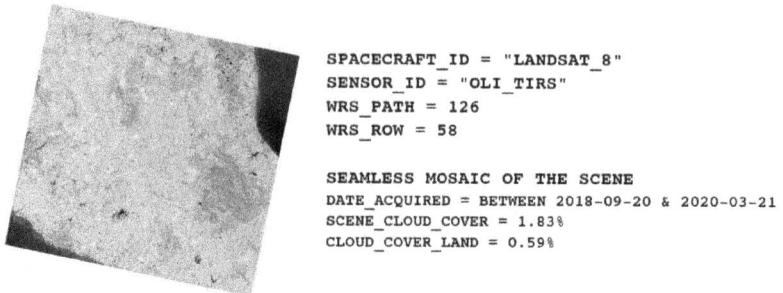

Figure 4.
Cloudless image of Landsat scene 126/058.

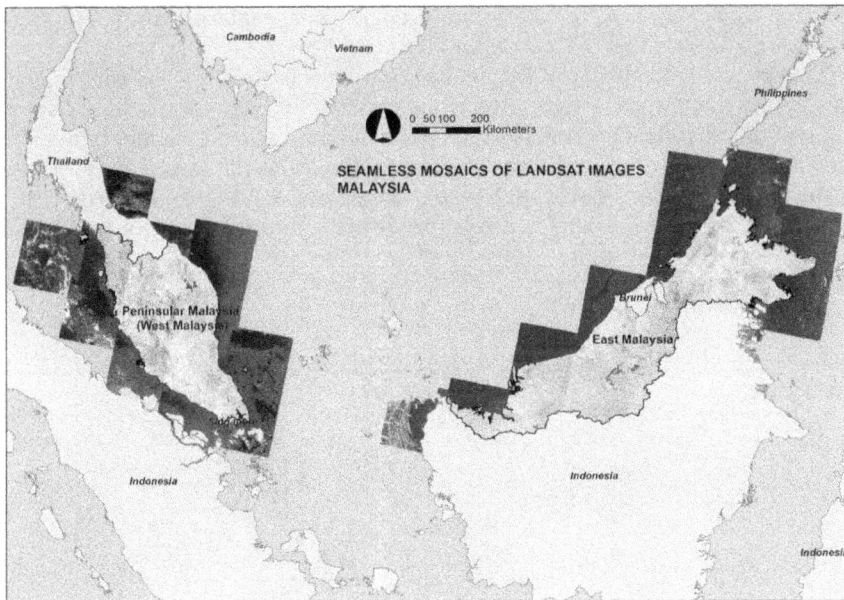

Figure 5.
Cloudless image of Landsat images over Malaysia.

the images, especially at the highlands and top of mountains areas. This is due to the clouds that always there all the time regardless weather conditions and seasons.

2.3 Image classification

Forest includes all land with woody vegetation consistent with threshold (minimum mapping unit (MMU) is 0.5 ha, minimum crown cover is 30% or minimum height at maturity is 5 m) used to define forest land in the national statistic. It also includes system with vegetation structure that currently falls below threshold, but in situ could potentially reach the threshold value is expected to exceed (the threshold of forest land category is sub-divided at the national level into managed and unmanaged and by ecosystem type as specified in the IPCC Guidelines) [14]. In this study, forests are divided into three major ecosystem types, which are inland forest, peat swamp forest and mangrove forest. These areas were further divided into Permanent Reserved Forests (PRFs)/Permanent Forest Estate (PFEs)/Permanent Forest Reserves (PFR) as managed category and the remaining areas outside the managed areas as stateland forest [15]. **Figure 6** illustrates how the forests is defined and various conditions (due to management practices and natural disturbances) that possibly occur in the forests in Malaysia.

Understanding these conditions and management practices in forestry sector in Malaysia are crucial before these forests are interpreted and classified on the satellite images. Having several secondary data before hands are desirable and can facilitated the classification processes. Spatial information such as boundary of the PRFs, the management regimes, types and locations of varying ecosystems are required to ensure that the classification is performed accurately. In this case, the classification was performed to delineate forests from other land features. This process was performed by using traditional supervised classification method. Several training sets were selected on the images. Unchanged forest areas, which were determined from secondary spatial data were used as forest training sets, and the other land cover classes were determined from the image interpretation. The same training set was used for forest class for all time series.

The biggest challenge in the image classification was about to deal with huge data size and to produce classification results with minimal uncertainties. Manual editing of the classification results was typical and need to be done repeatedly, which is a tedious process and time consuming. However, the results are satisfying, and the example of the classification results are depicted in **Figure 7** for the year 2020. The classification results in pixels form were converted to vector format

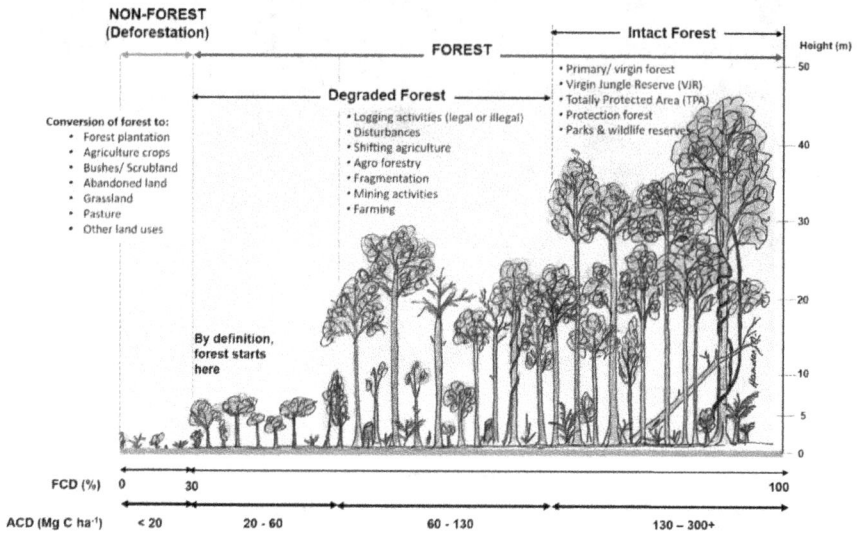

Figure 6.
Common structure of forests in Malaysia.

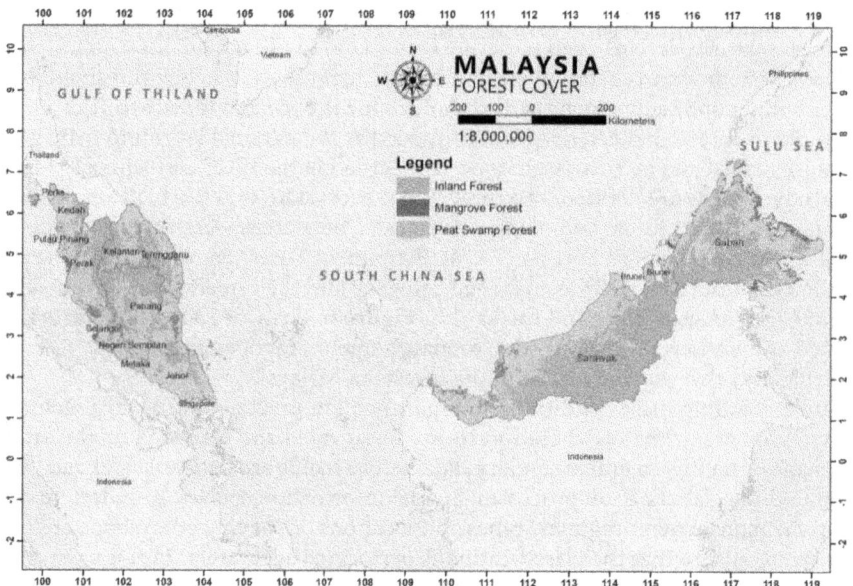

Figure 7.
Forests in Malaysia classified from the images.

Region	Forest Cover (ha)			Total Forest Cover (ha) $(d) = (a) + (b) + (c)$	Land Area* (ha) (e)	Percentage (%) (f) = (d)/ (e)*100
	Inland Forest (a)	Peat Swamp Forest (b)	Mangrove Forest (c)			
Peninsular Malaysia	5,338,082	243,504	110,953	5,692,539	13,100,367	43.5
Sarawak	7,328,029	320,207	139,890	7,788,126	12,444,951	62.6
Sabah	4,273,536	97,276	378,195	4,749,007	7,390,224	64.3
Total	**16,939,647**	**660,987**	**629,038**	**18,229,673**	**32,935,542**	55.3

*Sources: Department of Survey and Mapping Malaysia, Lands and Surveys Department, Sabah and Department of Land and Survey, Sarawak.

Table 4.
Composition of forest cover in Malaysia (2020).

(.shp) for further analysis and post-classification detection process. From the vector data, the areas for each forest type classified from the images were determined. Example forest area statistics derived from the vector data is summarized in **Table 4** for the year 2020.

2.4 CO_2 emissions calculation

Carbon dioxide (CO_2) is the main greenhouse gas that plays critical roles in regulating the earth's climate. According to IPCC, there are two basic approach to estimate CO_2 emissions/removals, which is Gain-Loss Method (GLM) and Stock-Difference Method (SDM). Calculation methods for this study are determined by SDM at Tier-2 level by using CO_2 based on [16]. The result is then multiplied by 44/ 12 or equal to 3.67 unit of carbon (C). Since the emission from the forestry activities are considered, the CO_2 in this study is attributed only from the forest carbon stock, and it is not equivalent to emission from other gases. Therefore, the reported emission is in carbon dioxide (CO_2) and not carbon dioxide equivalent (CO_2e).

2006 IPCC Guidelines offer a default methodology that includes default emission factors for Tier-1 [14]. Tier-1 level is designed to be the simplest to use, for which equations and default parameter values (e.g., emission and stock change factors) are provided by 2006 IPCC Guidelines. The emission factor is derived from readily available statistical information, which often globally available sources of activity data estimates (e.g., deforestation rates, global forest cover maps, etc.) although these data are usually spatially coarse.

Meanwhile, Tier-2 level use the same or similar activity data to Tier-1 level but applies emission and stock change factors that are based on country- or region-specific data. Country-defined emission factors are more appropriate for the local climatic regions and land use system. In many cases the Tier-2 could also be applied at a higher level of temporal and spatial resolution and more disaggregated activity data, where the activity statistics are further split into sub-categories.

Higher-order approaches are utilized at the Tier-3 level, such as models and inventory measurement systems suited to national circumstances, repeated over time, and driven by high-resolution activity data disaggregated at the subnational level. Higher-order approaches produce more accurate estimations than lower-tier approaches.

Estimated carbon stock (Mg C) in the stock change method is obtained by multiplying the forest area (ha) by the carbon stock per unit area (Mg C ha^{-1}). The carbon stocks of the entire project area at a given time are obtained by calculating the products of the carbon stocks per unit area for each forest type and the area occupied by that type and then summing the results over all forest types.

$$C_t = \sum_{i=1}^{n}(A_i \times C_i)$$ (1)

Where:
C_t = total carbon stock at a certain time t (Mg C)
A_i = area occupied by forest type i (ha)
C_i = carbon stock per unit area of type i (Mg C ha^{-1})
The emission is calculated as the difference of carbon stocks for a given forest area at two points of time, which is expressed as

$$\Delta C = (C_{t1} \times C_{t2})/(t_2 - t_1)$$ (2)

Where:
ΔC = annual carbon stock change in biomass (Mg C yr.$^{-1}$)
C_{t1} = carbon stock at time 1 (Mg C)
C_{t2} = carbon stock at time 2 (Mg C)

3. Results and discussion

3.1 Changes of forest cover

Deforestation is defined as human induced permanent conversion of forest land to non-forest, i.e., all the forest stands are cut, and the land is cleared and used for another purpose. Temporary change in land use, like one rotation tree crop (up to 25 years) within forest reserves are not considered as deforestation [17]. In a broader term, deforestation converts forest land to alternative, permanent, non-forested land to be used in agriculture, grazing or urban development or clearing of any area of its natural vegetation cover, which normally leads to a decrease in plant population, resulting in a loss of plant biodiversity [18]. Deforestation is caused by multiple drivers and pressures, including conversion for agricultural uses, infra-structure development, wood extraction, agricultural product prices, and a complex set of additional instructional and location-specific factors [7], which can be extremely important in certain localities.

A crude estimate showed that the total forest loss in Malaysia during years 2000–2012 amounted to 14.4% of its year 2000 forest cover [19]. Oil palm expansion was the major reason that contributed to the figure. The oil palm plantation area in Malaysia increased from 5.59 to 11.56 Mil. ha from 2000 to 2018, an increase of 5.98 Mil ha with a growth rate of 106.96%. The area of oil palm plantations in West Malaysia increased by 2.53 Mil ha, with a growth rate of 82.77%; in East Malaysia, the area increased by 3.45 Mil ha, with a growth rate of 136.14% [20]. The growth of oil palm accelerated between years 2000–2010 and become decelerated starting from 2010 onwards. In addition to that, the deforestation was caused by rubber plantation, construction of hydro-electric dam reservoirs, mining activities, forest fire, illegal logging, shifting cultivation, and natural disasters such as tsunami and erosion.

In contrast, the study found that the deforestation from 2005 to 2020 was amounted to the loss only of 1,087,030 ha (5.6%) of its year 2005 forest cover, with the annual rate of deforestation at 0.37% yr.$^{-1}$. Hence the study proved that the reported rate by [19] was not right. The forest cover has reduced from 19,316,702 ha in 2005 to 18,229,672 ha in year 2020 (**Table 5**). This was attributed to reduction in about 3.4% of total forest extents in Malaysia due to the conversion of forest to agricultural lands and settlement, which were mainly under the stateland forest that are designated for development purposes.

3.2 Carbon stock of forests in Malaysia

Aboveground biomass (AGB) comprises all living aboveground vegetation including stems, branches, twigs, and leaves. It is the most important pool of carbon forest types. In this study, a published allometric equation was used to calculate AGB for inland forests [21]. This equation was calibrated based on trees sampled in lowland and hill forests in west Peninsular Malaysia. Wood densities were obtained from the Global Wood Density Database [22]. A biomass expansion factor of 0.47 was used to convert the biomass into carbon stock. Previous study indicated that the average values for carbon stock from all carbon pools in major types of forest in Malaysia as summarized in **Table 6** [16]. A comprehensive review of carbon stock in various forest types and conditions in Malaysia was also made by [23, 24]. However, only the aboveground component of carbon stock is used for the emission calculation in this study.

The most important parameters that play roles that produce variations in carbon stock estimations are (i) the use of different allometric equations in the estimations, (ii) application of different sampling design/protocols, (iii) levels of disturbances in the forest, (iv) harvesting/ logging practices in production forest, and (iv) the selection of study sites. These influence the process of selecting project sites for carbon offset project.

Year	Forest cover (ha)				Percentage cover (%)
	Inland forest	Peat swamp forest	Mangrove forest	Total	
2005	17,949,753	700,401	666,547	19,316,702	58.7
2010	17,329,165	676,186	643,502	18,648,853	56.6
2015	17,088,338	666,789	634,559	18,389,686	55.8
2020	16,939,647	660,987	629,038	18,229,672	55.3

Given the landmass of Malaysia was at 32,935,542 ha.
Source: https://www.data.gov.my/data/ms_MY/dataset/keluasan-malaysia.

Table 5.
Forest cover in Malaysia (ha).

Forest type	Carbon stock (Mg C ha^{-1})					
	Above-ground	Below-ground	Dead-wood	Litter	Soil	Total
Inland forest	174.49	35.22	4.92	1.29	48.40	264.32
Peat swamp forest	168.63	35.95	21.40	2.19	188.10	416.27
Mangrove forest	135.45	48.57	22.12	3.88	54.87	264.89

Table 6.
Carbon stock in all carbon pools in major types of forests in Malaysia (Mg C ha^{-1}).

3.3 CO_2 emission from deforestation

Assuming the CO_2 emission occurred was a result from the changes of forest cover, the emission from year 2005 to 2020 was about 689.26 Mil. Mg CO_2, with an annual rate of emission at 45.95 Mil. Mg CO_2 yr.$^{-1}$. This was equal to carbon loss of about 12.53 Mil. Mg C. **Table** 7 summarizes the trend of CO_2 emission that occurred between years 2005 and 2020. The trend indicates that the deforestation accelerated between years 2005 and 2010 and slowed down between year 2010 and 2020. This was mainly due to awareness and mitigation action among government towards REDD+ interventions and enhancement of management practices towards various conservation efforts.

Although there are a few assumptions and generalizations were included in the estimations, the reported figures can present an overall scenario of CO_2 emission resulted from deforestation activities in Malaysia.

3.4 CO_2 emission in forest land remaining forest land

Although deforestation attributed much to the CO_2 emission in LULUCF sector, the remaining forests are still playing roles in CO_2 sequestration while they regrow. However, the rate of CO_2 sequestration is very slow and is greatly depending on the overall management practices applied within the forests. This is also typically occurred within the PRFs where some areas are designated for production purpose with sustainable forest management (SFM) practices. The average rate of seques-tration for the major types of forests in Malaysia is summarized in **Table 8**.

Analysis of CO_2 emission and removals from LULUCF sector in Malaysia evidenced that the activity data used are very important to determine the emission and removals within the forest land remaining forest land category. In this case, data such as logging history records, net production (timber volume), and annual allowable coupe (AAC) acquired from the respective forestry departments were used estimate emission from upstream forest operations. Peninsular has contributed net removal from the category was about -0.14 Mil. Mg CO_2 within 5 years from 2005 to 2010. Then, it was followed by net emissions at 14.31 Mil. Mg CO_2 which occurred between year 2010 to 2015 as compared to year 2015 to 2020, which

Time Series	CO$_2$ emission (Mil. Mg CO$_2$)			Total
	Inland forest	Peat swamp forest	Mangrove forest	
2005–2010	397.05	14.97	11.45	423.47
2010–2015	154.08	5.81	4.44	164.33
2015–2020	95.13	3.59	2.74	101.46

Table 7.
CO$_2$ emission resulted from deforestation Malaysia (2005–2020).

Forest type	Growth rate AGB (Mg ha^{-1} yr.$^{-1}$)	Carbon sequestration* (Mg C ha^{-1} yr.$^{-1}$)
Inland	9.3	4.37
Peat swamp	9.2	4.23
Mangrove	11	5.17

Carbon conversion factor: 0.47.

Table 8.
Rate of carbon sequestration in major forest types in Malaysia [17].

Year	Net (Mil. Mg CO₂)			
	Peninsular Malaysia	**Sabah**	**Sarawak**	**Total (entire Malaysia)**
2005–2010	−0.14	−16.22	−11.83	−28.19
2010–2015	14.31	−0.49	−90.44	−76.62
2015–2020	23.53	−72.62	48.87	−0.22

Note: -ve sign is net removal and + ve sign is net emission.

Table 9.
Summary of CO_2 emission for forest land remaining forest land.

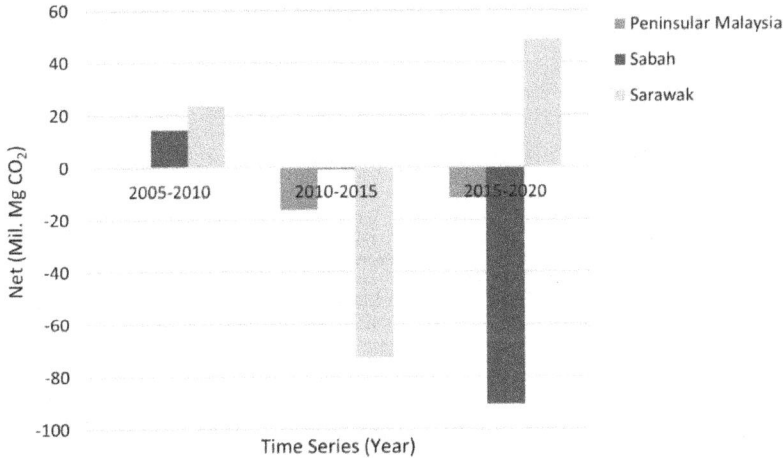

Figure 8.
Net CO_2 emission/removal in Malaysia.

accounted net emissions for 23.53 Mil. Mg CO₂. Meanwhile, Sabah has contributed net removal from forest land remaining forest land about −16.22 Mil. Mg CO₂ within 5 years from 2005 to 2010. Then, it continued to remove emission at −0.49 Mil. Mg CO₂ between years 2010 and 2015, and even much greater between years 2015 and 2020, with the accounted net removal of −72.62 Mil. Mg CO₂. Sarawak has contributed net removal of about −11.83 Mil. Mg CO₂ in 5 years between 2005 and 2010 and continued to the years between 2010 and 2015 with a net removal at −90.44 Mil. Mg CO₂. However, it emitted back within the years 2015–2020 with the net emission of 48.87 Mil. Mg CO₂. **Table 9** and **Figure 8** summarize the fluctuations in the net emission and removal that have occurred within years 2005 to 2020.

Principally there is not fix trend in the net emission and removal within the category of forest land remaining forest land in each region in Malaysia. This indicates that the activities within the forests are dynamic and unpredictable. Some areas could produce emission, but some other areas sequester carbon and thus resulting in removals. However, taking Malaysia as a whole, there is a trend of continuous removals from year 2005 to 2020 that has been produced by the remaining forests in Malaysia (**Figure 9**).

4. Requirements to offset the emission

It is concluded that the rapid logging operations within the forest can be very dynamic, thus the forests in Malaysia not only remove CO₂ but also produce

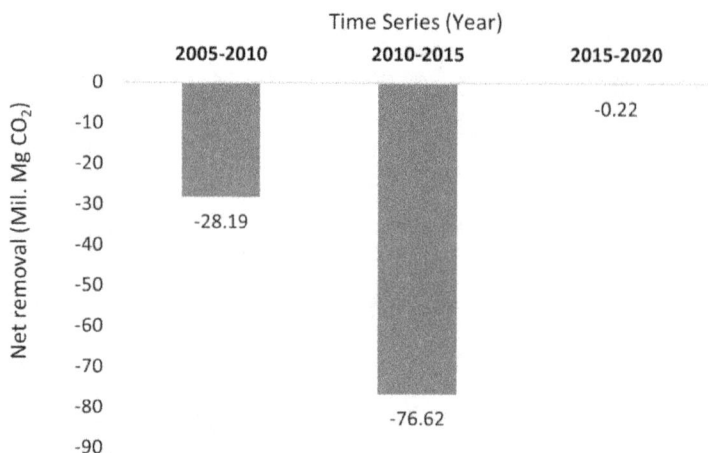

Figure 9.
Net CO₂ removal in Malaysia.

Time Series	Net CO2 emission/removal (Mil. Mg CO₂)		
	Net CO₂ emission from deforestation	Net CO₂ removal in forest land remaining forest land	Net CO₂ emission
2005–2010	423.47	−28.19	395.28
2010–2015	164.33	−76.62	87.71
2015–2020	101.46	−0.22	101.24

Table 10.
Summary of CO₂ emission for forest land remaining forest land.

emissions. These indicated that the forest areas that have been serving for timber production and logging are continuously fluctuating in terms of emission and removals. This area is worth for a carbon offset program because the intervention can stand on all pillars of SFM; ecological, economic, and socio-cultural. The forests' ability to attract investment and support commercially sustainable forest uses is unaffected in the present and future.

Analysis indicated that the LULUCF sector in Malaysia is still producing the net emission at 101.24 Mil. Mg CO₂ at the end series of year 2020 (**Table 10**). Based on this figure, Malaysia needs to stop completely deforestation activities and restore about 27.6 Mil. Mg C, which is equal to 158,240 ha of natural forest to offset the emission from LULUCF sector in the country. Otherwise, Malaysia must limit logging activities and retain about 631,838 ha of the logging areas to regrow naturally for at least 10 years. This is almost impossible since Malaysia is a developing country and still depending much on the forests for timber productions [25]. While the remaining stateland forests can be demolished at any time for development purposes. Therefore, the finding of this study suggesting that the project types that have potentially suitable for carbon offset program are (i) avoided planned deforestation, (ii) conversion of logged forests to protected forests, (iii) extending the rotation age of evenly aged managed forests, and (iv) avoided deforestation on wetlands (conservation).

The continuing removals indicated that the forest sector contributed greatly to the sinks through forest land remaining forest land. In this case, there are activities occurred within the forestland that attributed to the CO_2 removals, which could be driven by the good of management practices, such as Forest Stewardship Council (FSC) Certification Scheme. The FSC will conduct an independent review of forest management methods in order to ensure sustainable management practises, sustainable management of Malaysia's natural forest, and to meet demand for certified timber products.

Forest enrichment activities are being made by the forest industry between 2016 and 2020 to rehabilitate degraded forests. Ongoing projects such as the Central Forest Spine (CFS) in Peninsular Malaysia and the Heart of Borneo (HoB) in Sabah and Sarawak serve as facilitators for improving forest connectivity, reducing fragmentation, and improving natural resource management. The forestry sector implemented a REDD+ strategy in 2017 to ensure that at least 50% of Malaysia's land mass is forested, which was accomplished by improving sustainable forest management, conservation activities, and seeking synergies with activities under the National Policy on Biological Diversity 2016–2025 [17].

Nonetheless, the current net removals are still not sufficient to offset the emission that has been produced by the deforestation activities in Malaysia. While the LULUCF sector is producing emission, the other sectors such as energy, transportation, agriculture, solid waste, and others are also emitting CO_2 to the atmosphere at even greater amount. Therefore, carbon offset can not only depend on forests. Appropriate mitigation actions need to put in proper place within the individual sector so that the climate change mitigation can be achieved, and the targeted reduction of global temperature is materialized.

5. Conclusion

The study demonstrated that the use of remote sensing data, coupled with the other supporting data are viable for assessing forest carbon and emissions in forestry sector in Malaysia. Although there were technical issues regarding the data, with appropriate image processing methods, the issues have been well addressed. Landsat satellite images that have been acquired between years 2005 and 2020 with 5-year intervals were processed to produce seamless, wall to wall images over Malaysia. Forests have been identified from the image classification and then classified into three major types, which are dry-inland forest, peat swamp and mangroves. Post-classification change detection technique was used to determine areas that have been undergoing conversions from forests to other land uses.

Forest areas were found to have declined from about 19.3 Mil. ha (in 2005) to 18.2 Mil. ha in year 2020. The study found that the deforestation from 2005 to 2020 was amounted to the loss of 1,087,030 ha (5.6%) of its year 2005 forest cover, with the annual rate of deforestation at 0.37% yr.$^{-1}$. This has contributed to the total CO_2 emission of 101.46 Mil. Mg CO_2. The study also estimated the total CO_2 emission and removals within the forest land remaining forest land. It was revealed that the forests also produced emission in terms of timber production activities. However, the overall estimates showed that this category is still able to sequester carbon and provide removals at a sum of 105.03 Mil. Mg CO_2 for the period of 15 years (2005–2020).

The study exposed suggested that Malaysia must stop completely deforestation activities and restore about 27.6 Mil. Mg C to achieve the net-zero emission. This is equal to 158,240 ha of natural forest or 631,838 ha of the logging areas to need to be left regrown naturally for at least 10 years. The study also suggested that the project types that have potentially suitable for carbon offset program in Malaysia are (i) avoided planned deforestation, (ii) conversion of logged forests to protected forests, (iii) extending the rotation age of evenly aged managed forests, and (iv) avoided deforestation on wetlands (conservation).

The study proved that the use of a series satellite images from optical sensors are the most appropriate sensors to be used for monitoring deforestation in Malaysia. Although cloud covers are the major issue for optical imagery datasets, current development in remote sensing, computer technologies and processing algorithms for images analysis can provide solutions for the issues.

Acknowledgements

This study was conducted in collaboration between Geoinformation Programme, Forest Research Institute Malaysia (FRIM) and National Petroleum Limited (PETRONAS) under the project entitled "Screening Study for Carbon Offset Opportunities in Malaysia". Thanks to Forestry Department Peninsular Malaysia (FDPM), Sabah Forestry Department (SFD) and Forest Department of Sarawak (FDS) for providing supporting related data and statistics that was used for the calculations in this study. The publication of this chapter is supported by the FRIM's research and development (R&D) fund under the 12th Malaysia Plan from the Government of Malaysia through the Ministry of Energy and Natural Resources, Malaysia. Thanks also to the USGS (https://earthexplorer.usgs.gov) that provide free-access Landsat images for this study.

Conflict of interest

The authors declare no conflict of interest.

Author details

Hamdan Omar[1*], Thirupathi Rao Narayanamoorthy[2],
Norsheilla Mohd Johan Chuah[1], Nur Atikah Abu Bakar[2]
and Muhamad Afizzul Misman[1]

1 Forest Research Institute Malaysia (FRIM), Kepong, Selangor, Malaysia

2 National Petroleum Limited (PETRONAS) Petroliam Tower 1, Petronas Towers, Kuala Lumpur Convention Centre, Kuala Lumpur, Malaysia

*Address all correspondence to: hamdanomar@frim.gov.my

IntechOpen

References

[1] Brendan M, Cyril FK, Heather K, William RM, Houghton RA., Russell AM, Hole D, Sonia H. Understanding the importance of primary tropical forest protection as a mitigation strategy. Mitigation and Adaptation Strategies for Global Change, 2020. 25: 763–787.

[2] Lu D. The potential and challenge of remote sensing-based biomass estimation. International Journal of Remote Sensing 2006. 27(7): 1297-1328.

[3] Nurul Ain MZ, Zulkiflee AL. Carbon sinks and tropical forest biomass estimation: a review on role of remote sensing in aboveground-biomass modelling. Geocarto International, 2017. 32(7): 701-716.

[4] Hamdan O, Muhamad Afizzul M, Abd Rahman K. Synergetic of PALSAR-2 and Sentinel-1A SAR Polarimetry for Retrieving Aboveground Biomass in Dipterocarp Forest of Malaysia. Applied Sciences. 2017. 7, 675.

[5] Pearson TR, Brown S, Murray L, Sidman G. Greenhouse gas emissions from tropical forest degradation: an underestimated source. Carbon Balance and Management, 2017. 12, 3.

[6] IPCC. Global Warming of 1.5°C. An IPCC Special Report [Internet]. 2018 Available from: https://www.ipcc.ch/sr15/ [Accessed: 2020-07-24].

[7] UNFCCC. Malaysia's Submission on Reference Levels for REDD+. [Internet]. 2020. Available from: https://www4.unfccc.int/sites/submissions/INDC [Accessed: 2020-02-12].

[8] Ministry of Natural Resources and Environment Malaysia (NRE). A Roadmap of Emissions Intensity Reduction in Malaysia. 2014. Putrajaya, Malaysia.

[9] Lopez G, Laan T. Biofuels—At what cost? Government Support for Biodiesel

in Malaysia; International Institute for Sustainable Development: Geneva, Switzerland [Internet]. 2008. Available from: https://www.iisd.org/gsi/sites/default/files/Final_Malaysia_2.pdf [Accessed: 2020-06-03]

[10] Verified Carbon Standard (VCS). The VCS Program. [Internet]. 2020. Available from: https://www.offsetguide.org/ [Accessed: 2020-04-20].

[11] Chris W. Carbon offsets. [Internet]. 2016. Available from: https://www.explainthatstuff.com/carbonoffsets [Accessed: 2020-02-22].

[12] Crouzeilles R, Alexandre N, Bodin B, Beyer HL, Guariguata MR, Chazdon RL. How to Deliver Forest Restoration at Scale: Recommendations for unlocking the potential of the most cost-effective way to restore forests in the fight against climate change and biodiversity loss. Conservation Letters. 2019.

[13] Qiu S, Zhu Z, He B. Fmask 4.0: Improved cloud and cloud shadow detection in Landsats 4–8 and Sentinel-2 imagery. Remote Sensing of Environment, 2019. 231: 111205.

[14] IPCC. 2019 Refinement to the 2006 IPCC Guidelines for National Greenhouse Gas Inventories. [Internet]. 2019. Available from: https://www.ipcc.ch/report/2019-refinement-to-the-2006-ipcc-guidelines-for-national-greenhouse-gas-inventories/ [Accessed: 2020-08-16].

[15] Ministry of Natural Resources and Environment Malaysia (NRE). National REDD Plus Strategy. 2018. Putrajaya, Malaysia.

[16] Hamdan O, Norsheilla MJC, Ismail P, Samsudin M, Wan Abdul Hamid Shukri WAR, Azmer M. Forest

Reference Emission Level for REDD+ in
Pahang, Malaysia. FRIM Research
Pamphlet No. 141, 2018. 97 pp.

[17] Ministry of Energy, Science,
Technology, Environment and Climate
Change (MESTECC). Malaysia, Third
National Communication and Second
Biennial Update Report to the UNFCCC.
2018. Putrajaya, Malaysia.

[18] Omran A, Schwarz-Herion O.
Deforestation in Malaysia: The Current
Practice and the Way Forward. In:
Omran A., Schwarz-Herion O. (eds)
Sustaining our Environment for Better
Future. 2020. Springer, Singapore.

[19] Butler R. Malaysia has the world's
highest deforestation rate, reveals
Google forest map. [Internet]. 2013.
Available from: http://news.mongabay.c
om/2013/1115-worlds-highest-deforesta
tion-rate [Accessed: 2020-07-21].

[20] Li W, Fu D, Su F, Xiao Y. Spatial–
Temporal Evolution and Analysis of the
Driving Force of Oil Palm Patterns in
Malaysia from 2000 to 2018. ISPRS Int.
J. Geo-Inf. 2020. 9: 280.

[21] Chave J, Réjou-Méchain M,
Búrquez A. et al. Improved allometric
models to estimate the aboveground
biomass of tropical trees. Glob Change
Biol, 2014. 20: 3177-3190. doi:10.1111/
gcb.12629

[22] Chave J, Andalo C, Brown S. et al.
Tree allometry and improved estimation
of carbon stocks and balance in tropical
forests. Oecologia, 2005. 145, 87–99.

[23] Kho LK, Jepsen MR. Carbon stock of
oil palm plantations and tropical forests
in Malaysia: A review. Singapore Journal
of Tropical Geography, 2015. 36: 249–
266.

[24] Syafinie AM, Ainuddin AN.
Aboveground Biomass and Carbon
Stock Estimation in Logged-Over
Lowland Tropical Forest in Malaysia.

International Journal of Agriculture,
Forestry and Plantation, 2015. 1: 1- 14.

[25] Deutch J. Is Net Zero Carbon 2050
Possible? Joule, 2020. 4: 2237–2243.

Chapter 4

Optical Remote Sensing of Planetary Space Environment

Fei He, Zhonghua Yao and Yong Wei

Abstract

Planetary science is the scientific investigations of the basic characteristics and the formation and evolution processes of the planets, moons, comets, asteroids and other minor bodies of the solar system, the exoplanets, and the planetary systems. Planetary scientific research mainly depends on deep space exploration, and it is highly interdisplinary and is built from Earth science, space science, astronomy and other relevant disciplines. Planetary space, a critical region of mass and energy exchange between the planet and the interplanetary space, is an integral part of the planetary multi-layer coupling system. Atmospheres of different compositions and plasmas of different densities and energies exist in planetary space, where mass transportation at different temporal and spatial scales and various energy deposition and dissipation processes occur. Optical remote sensing overcomes the difficulties of capturing global views and distinguishing spatiotemporal variations in in-situ particle and field detections. This chapter introduces the principles and applications of optical remote sensing in planetary science. The first ground-based planetary observatory in China, the Lenghu Observation Center for Planetary Sciences, will be introduced in detail. Future development of optical remote sensing platforms in Chinese planetary exploration program will also be introduced.

Keywords: Planets, Space Environment, Optical Remote Sensing, Atmosphere, Space Plasma, Radiation Mechanisms

1. Introduction

How a planet is formed? What initial conditions and combinations of subsequent geological, chemical, and biological processes lead to at least one planet that is the home of innumerable life forms? What determines the fate of lives on a planet? These scientific questions can be summarized to three fundamental questions that human beings are always thinking about: where we come from, how we develop to current state, and where we will go? These questions are closely related to science, religion, philosophy, humanity, and other fields.

Thousands of years ago, or in the much more distant past, the human beings have already started looking at the stary sky to try to understand how everything works and which corner our planet is at in the Universe. However, limitations of the eyes also confined the thoughts of human beings, until the 1910s, G. Galilei developed the first astronomical telescope and use it to observe the Universe. With the help of optical telescope, our field of vision is greatly extended to the deep Universe. Discoveries by G. Galilei, such as the four Galilean moons circling Jupiter, the variation of Venus phase, and the sunspot, opened a new era of planetary science and astronomy. Observation and exploration of planets transformed human's philosophical thinking to scientific activations, deeply changed the human's path to the answer.

In 1668, I. Newton developed the first reflective optical telescope, also called Newton telescope, and a window was opened for large aperture telescopes, based on which, we could see the Universe further and clearer. In 1789, the British scientist F. W. Herschel developed the first reflective optical telescope with aperture larger than one meter. Using this large telescope, he discovered Uranus and its two moons. As the increase of our requirements on observing further and weaker targets, the aperture and optical performance of the telescope become the main limitations of the applications of optical remote sensing in planetary science. Until the middle of the 20th century, benefited from the breakthroughs in optical techniques, such as the fabrication of high-accuracy large aperture mirrors, the maturity of optical aberration correction technology, the development of opto-electronic detectors, and the applications of active optics and adaptive optics, the ground-based large aperture optical telescopes have greatly advanced, and many new results were obtained. For example, the discovery of Na and K in Mercury atmosphere, the discovery of large amount of CO_2 in the Venusian atmosphere, the discoveries of CO_2, H_2O, and CH_4 in Martian atmosphere, the discovery of neutral nebula and Io plasma torus around Jupiter.

Nowadays, even the aperture of the ground-based optical telescope has achieved tens of meters (e.g., the Extremely Large Telescope currently under construction by European Southern Observatory), our vision is still limited by the atmospheric envelope. First, due to the absorption and attenuation of the atmospheric gases, observations at many wavelengths are unavailable, such as wavelengths shorter than ultraviolet (UV) and specific absorption bands in infrared (such as H_2O, O_3, CO_2, CH_4). Second, atmospheric turbulence greatly limited the resolution of the telescopes and diffraction-limited imaging is difficult to be achieved even with adaptive optics. Third, the atmospheric background emissions limited the observable time and the detectable weakest emission. The best way to escape from the constraints of the atmosphere is to go above the atmosphere, e.g., in the stratosphere or in the space.

In the 1960s, the stratospheric balloons have been used to observe planetary atmospheres. When a balloon floats above 35 km from the surface, it rises above 99.5% of the atmosphere, all of the telluric water vapor, and almost all of the CO_2 and CH_4, the radiations at the wavelengths from near-UV (200–400 nm) to far infrared (tens of micrometers) become detectable, and the conditions are like space in some regards. Moreover, influence of atmospheric turbulence becomes negligible, diffraction-limited imaging can be utilized for telescopes of several meter aperture, allowing for a marked improvement in the observation resolution. Although the daytime sky brightness is much brighter in the stratosphere but may still allow daytime observations, particularly at long wavelengths and at angles away from the Sun. In 1964, Bottema et al. determined the amount of water vapor presented above the reflective cloud layer on the Venus using an automatic daytime telescope of 30-cm aperture carried by balloon to 26.5 km. Currently, more and more attentions have been paid to balloon-borne planetary explorations, for example, the Scientific Experimental system in Near SpacE (SENSE) program in China, the European Stratospheric Balloon Observatory (ESBO) program in Europe, the Fujin program in Japan, the Balloon Rapid Response for ISON (BRRISON) mission in the United States.

The development of space technologies since the 1950s completely set the optical remote sensing free in space and the scientists and engineers have taken the remote sensing play the most incisive. Theoretically, all the wavelengths can be detected, and high-resolution diffraction-limited imaging can be realized. Compared with traditional in-situ particle and field detections, optical remote sensing is sensitive in characterizing matter compositions and overcomes the difficulties of capturing global views and distinguishing spatiotemporal variations. Therefore, optical payloads including photometer, spectrometer, spectrograph, and imager have been widely carried on almost all the planetary spacecraft. These different types of instruments greatly helped us uncover the secrets of planets that are shaded by the terrestrial atmosphere.

Benefited from the first upsurge of the international deep space exploration in the 1960s–1970s, a new and interdisciplinary discipline, the planetary science, is established developed in western countries. Planetary science is the scientific investigations of the basic characteristics and the formation and evolution processes of the planets, moons, comets, asteroids and other minor bodies of the solar system, the exoplanets, and the planetary systems [1]. A planet is a multilayer coupling system that contains (not completely and not limited) interior, surface, atmosphere and space. The Earth is the planet with the most complete spheric layers and is the only planet with life in the solar system. The interior and geological processes generate the atmosphere above the planetary surface and extends to space.

In this chapter, logical understanding of principles and applications of optical remote sensing in planetary science is delivered. The planetary space environment will be briefly introduced in Section 2. Then, the principles of optical remote sensing and the planetary optical radiations will be introduced in detail in Section 3, followed by current and future optical remote sensing plans in China in Section 4. Finally, a summary and outlook will be present in Section 5.

2. What is planetary space environment?

The space surrounding the Earth and other planets, the space between planets, and the space between stars and planets are far different from the environment we regularly experience on Earth. It is not empty and far from calm. It's a complex electromagnetic system in which variational magnetic fields generate electric currents and vice versa, while neutral or charged particles of different energy experience complex dynamics in the electromagnetic field. Particle and electromagnetic radiations continuously blowing from the sun (or a star), cosmic rays outside the solar system (or an extrasolar planetary system), and the planets themselves (e.g., planetary magnetic fields and even planetary weather) can cause changes in the system. Understanding the forces that drive the changes in space environment of our planet Earth and other planets not only help us protect technology and astronauts from radiation hazards, but also help us understand what makes a planet habitable. Incubation of lives on a planet takes more than just the right distance from a star. The star's interaction with the planet's atmosphere and electromagnetic system can make all the difference between a planet that's too dry, too hot, or too radiation-filled versus one where life could take root.

Generally, the space above the surface of a planet can be regarded as planetary space environment. The broad topic of physics of planetary space environment, designated planetary space physics, a subdiscipline in planetary science, focuses on the particles and fields within the space regions of the solar system and its immediate vicinity. For planets and moons, the space regions specifically refer to the neutral upper atmosphere, the ionosphere, the magnetosphere and the interplanetary space. Usually, planetary space research does not extend downward into the thick lower atmosphere of planets and moons, which is traditionally relegated to the realm of meteorology. Nevertheless, the planet is an integrated system, in which its multi-spheres are coupled, from the space to the inner core. Therefore, in the perspective of planets' evolution, we should treat the planets and their space environment as a whole system. Optical remote sensing can be applied to every aspect of this system, e.g., planetary geological activities, atmospheric activities and space plasma activities.

In the solar system, the planets can be divided into different categories according to variety criteria, for example, the terrestrial planets (Mercury, Venus, Earth and Mars) which are also called rocky planets, and Jovian planets (Jupiter, Saturn, Uranus, Neptune) which are also called gas giants. The planets can also be divided into two categories, one for magnetized planets with intrinsic dipolar magnetic field (Mercury, Earth, Jupiter, Saturn, Uranus, Neptune, and the largest moon of Jupiter, the Ganymede), the other for unmagnetized planets without intrinsic dipolar magnetic field (Venus and Mars). The space environments of magnetized planets, unmagnetized planets, asteroids and comets exhibit significant differences [2].

For the space environment of magnetized planets, the Earth has representativeness most. A schematic illustration of the terrestrial magnetosphere is shown in **Figure 1**. The magnetospheres of other planets with intrinsic magnetic field are similar but with different spatial scales decided by the magnetic field strength (**Table 1**, summarized from reference [2]). The magnetosphere, the region dominated by the planet's magnetic field, is a part of dynamic, interconnected system that responds to solar, planetary, and interstellar conditions. On the sun-facing side, or dayside, constant bombardment by the solar wind compresses the magnetic field and forms a magnetopause at a distance of about six to 10 times the radius of the Earth, depending on the activity of solar wind. On the nightside, the magnetosphere stretches out into an immense magnetotail, which can measure hundreds of Earth radii.

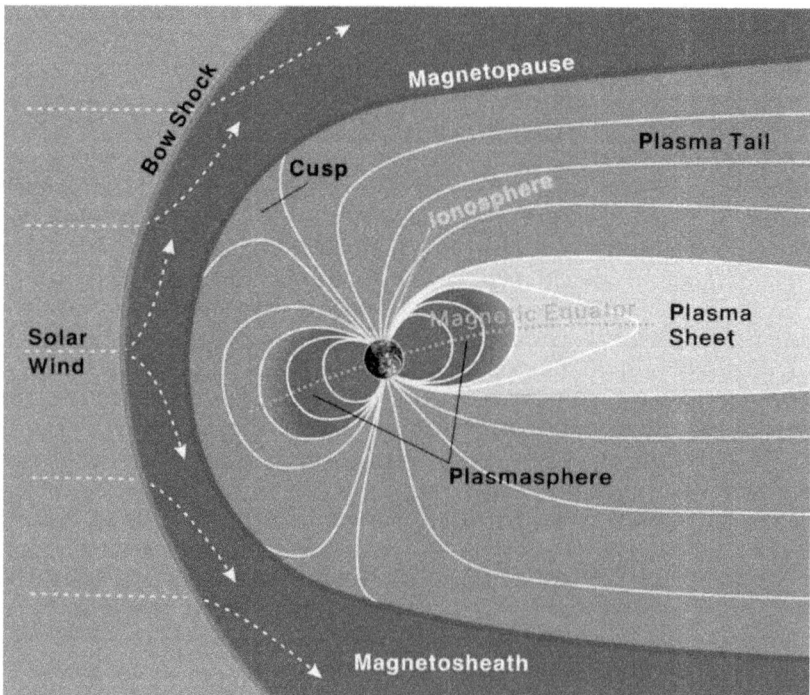

Figure 1.
Illustration of the Earth's space environment.

Parameters	Mercury	Venus	Earth	Mars	Jupiter	Saturn	Uranus	Neptune
Distance (AU)[a]	0.31–0.47	0.723	1	1.524	5.2	9.5	19	30
Radius, R_P (km)	2,439	6,051	6,373	3,390	71,398	60,330	25,559	24,764
Surface pressure (atm)[b]	<10–14	90	1	0.006	>>1000	>>1000	>>1000	>>1000
Magnetic moment (M_{Earth})[c]	4×10^{-4}	—	1	—	20,000	600	50	25
Surface magnetic field, B_0 (nT)	3×10^2	<2	3.1×10^4	<10	4.28×10^5	0.22×10^5	0.23×10^5	0.14×10^5
Solar wind density, ρ_{SW} (cm^{-3})	35–80	16	8	3.5	0.3	0.1	0.02	0.008
Magnetopause nose distance (R_{MP})[d]	1.4–1.6	—	10	—	42	19	25	24
Plasma density (cm^{-3})	~1	—	1-4000	—	>3000	~100	3	2
Composition	H+	—	O+/H+	—	O^{n}/S^{n+}	$O^+/H_2O^+/H^+$	H+	N+/H+
Dominant source	Solar wind	—	Ionosphere[e]	—	Io	Rings and moons[f]	Atmosphere	Triton
Time scale	Minutes	—	Days/Hours[e]	—	10 ~ 100 days	30 days–years	1–30 days	~1 day
Plasma motion	Solar wind drivent	—	Rotation/Convection[e]	—	Rotation	Rotation	Solar wind/rotation	Rotation

[a] $1\,AU = 1.5 \times 10^8\,km$.

[b] Refence to planetary fact sheet: https://nssdc.gsfc.nasa.gov/planetary/planetfact.html.

[c] Normalized to the magnetic moment of the Earth, $M_{Earth} = 7.906 \times 10^{15}\,T\,m^3$.

[d] $R_{MP} = (B2\,0/2\mu_0\rho_{SW}u^2)^{1/6}$ for typical solar wind conditions of ρ_{SW} given above and $u \sim 400\,km\,s^{-1}$. For outer planet magnetospheres, this is usually an underestimate of the actual distance.

[e] The dominant source inside the plasmapause is the ionosphere [3] and is mainly controlled by the corotation electric field with time scale of days. Outside the plasmapause, the dominant source is solar wind and is controlled by solar wind-driven convection electric field with time scale of hours.

[f] Enceladus, Tethys, and Dione.

Table 1.
Parameters of planetary magnetosphere in solar system.

The dipolar configuration of the Earth's magnetic field acts like a huge magnetic bottle, within which many charged particles are confined [2]. The motion of charged particles in the geomagnetic field is complicated, but can generally be divided into three types, namely gyration around a field line, bouncing between mirror points on the opposite side of the magnetic equator, and drifting around the Earth – westward for ions and eastward for electrons – due to the longitudinal gradient of the field lines. According to the energy and origin of the particles, the magnetosphere can be divided into several typical regions, namely, from inner to outer, the plasmasphere, radiation belt, ring current, plasma sheet, plasma tail in the case of Earth (**Figure 1**).

3. Optical remote sensing

3.1 Principle of optical remote sensing

Optical remote sensing is a remote detection method that studies an object through its optical emissions without coming into direct contact with it. Generally, optical remote sensing can generally be categorized into three types, i.e., imaging, spectrometry, and spectrographic imaging, as illustrated in **Figure** 2. The imaging, just likes we see the world through our eyes, directly project the radiations in the three-dimensional (3D) space to a two-dimensional (2D) detector. A filter with narrow band or wide band is used to select the wavelengths that can be recorded by the detector. The advantage of imaging is to separate the spatial and temporal variations in a certain time scale, and the spatial distribution of a specific composition that emit radiations at specific wavelength can be captured in a large scope of space. Therefore, imaging is the preferred method to investigate global spatiotemporal evolutions. The spectrometry, which split the incident radiations into single wavelengths through a dispersion element (e.g., prism) or an interferometer to measure the wavelength dependence of the incident radiations. Generally, such observation does not have spatial resolution and only single 'point' (Here, 'point' means a spatial region covered by the field of view of the spectrometer but not a geometrical point.) measurement can be realized, but the advantage is to identify different compositions or different

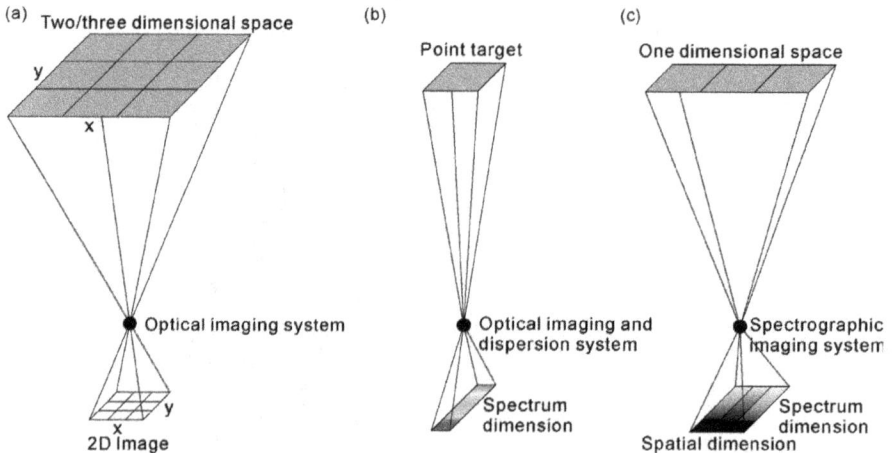

Figure 2.
Three typical methods of optical remote sensing (adapted from He, 2020). (a) Imaging. (b) Spectrometry. (c) Spectrographic imaging.

states (such as temperature and velocity) of a specific composition. Therefore, spectrometry is the primary measure to detect chemical composition and property in space. The spectrographic imaging, which integrated the advantages of imaging and spectrometry, can capture both spatial information (usually a line of view) and spectral information. Equipped with scanning mechanism, spatial coverage can be realized. In a global view, although the temporal resolution is reduced, the quasi-simultaneously captured spatial and spectral information is critical in global large-scale investigations.

The optical remote sensor is usually composed of an optical system that collect radiations and a detector that record the radiation through a photoelectric conversion, so we can analyze the digitalized signals to retrieve the physical information of the targets. The main performance indicators for an optical remote sensor include (not limited to the following indicators depending on different applications):

Field of View (FOV): the object space covered by the sensor. The larger the field of view, the larger spatial coverage. The spatial coverage can be calculated once the field of view and the distance to object are known.

Focal Length (f): the distance between the image point of infinity object and the principal plane of the optical system.

Operation Wavelength (λ): the wavelength range that the sensor can response. It is usually determined by a combination of the reflectivity of the mirrors, the transmission of the filters, and quantum efficiency of the detector. For spectrometric instrument, it depends also on the property of the dispersion elements (e.g., grating and prism).

Angular Resolution ($\Delta\theta$): the ability of the optical system to distinguish the image speckles of two objects. The diffraction-limited angular resolution $\Delta\theta$ is determined by the Rayleigh Criterion, $\Delta\theta=1.22\lambda/D$, where D is the aperture of entrance pupil.

Pixel Resolution (Δp): the angle corresponding to a pixel of the detector, $\Delta p = 2\tan(d/2f)$, where d is the size of the pixel and f is the focal length of the optical system. The angular resolution refers to the spatial resolving power of the optical system, while the pixel resolution is based on the physical pixel size of the detector. In actually optical design, the angular resolution and pixel resolution should be matched according to the radiation intensity and the requirement of signal-to-noise ratio (SNR). According to the Nyquist Criterion, two physical pixels are usually needed to resolve an optical resolution unit.

Area of Entrance Pupil (A): the area of the image of the aperture stop to the preceding optical system. The entrance pupil determines the size of the aperture of incident beam and thus determines the radiation energy that enter the optical system.

Spectral Resolution ($\Delta\lambda$): the ability to resolve radiations at different wavelengths.

Illumination Uniformity: the uniformity of the response of optical system (including the detector) to uniform incident radiations. Due to the aberration of optical system and manufacturing error of the optical elements, the detector response is nonuniform though out the image plane, and such nonuniformity should be measured during calibration process of the sensor. The calibrated matrix can be used to correct the scientific data.

Pixel Sensitivity (S): the ability of the response of the sensor to radiation intensity change. Commonly, the definition of S (in unit of count s^{-1} Rayleigh^{-1} pixel^{-1}) in space environment remote sensing is that, for an incident beam that fulfill entrance pupil with unit intensity (in Rayleigh, 1 Rayleigh = $10^6/4\pi$ photon cm^{-2} s^{-1} sr^{-1}), the count number or signal strength on the detector is $S(\lambda) = 10^6 A\Omega\eta(\lambda)/4\pi$, where A is the area of entrance pupil in cm^2, is the solid angle corresponding to a pixel in sr, and $\eta(\lambda)$ is the transmission efficiency of the system.

Exposure Time (ΔT): the integration time to acquire image with sufficient SNR.

Calibration Accuracy: the calibration of an optical sensor includes geometric calibration and radiometric calibration. The geometric error of an optical sensor originates from both the distortion of the optical system and the alignment error. The objective of geometric calibration is to accurately determine the projection relationship between pixel and geometric space. Usually, this can be achieved by imaging standard grid in laboratory. Radiometric calibration includes relative radiometric calibration and absolute radiometric calibration. The relative calibration is to measure the illumination uniformity of the sensor and the absolute calibration is to measure the pixel sensitivity of the sensor. Both are critical to retrieve the emission intensity of target from the optical images.

The above-mentioned indicators are common in general planetary optical remote sensing missions, some specific environmental adaptation indicators must be considered in specific application scenario, for example, the level of stray light, the ability of energetic particle shielding, and temperature dependence.

3.2 Optical radiation mechanisms

Optical emissions in space environment are the basis of remote sensing. Variations of the emissions reflect the physical property of the radiator. Understanding of the physical process of optical emission is developed after the establishment of quantum mechanics. The typical emission mechanisms for atoms, molecules, and ions are illustrated in **Figure 3**. The trigger of the emission is external energy, including optical radiation, particle collision, vibration and rotation (molecules and molecular ions), that excite the electron from ground state to higher energy state, the subsequent transition of electron to ground state emit a photon with the wavelength determined by the energy difference between the two states. The transition rate, intensity of spectral line, shifting of spectral line and broaden of spectral line can be determined by the density, temperature, and velocity of the radiator. It is noted that the mechanisms illustrated in **Figure 3a-h** mainly occur in the sunlit region. In the dark region, the emissions are primarily due to chemical reactions between different compositions. The collision charge-exchange process usually occurs in planetary magnetosheath region, between the energetic ions and low energy neutral atoms. For example, the collision between the hot solar wind charges particles (e.g., He^{2+}, O^{6+}) and the geocoronal neutral hydrogen generated EUV and soft X-ray emissions in the terrestrial magnetosheath and cusp regions [5], and in the dayside Martian ionosphere [6]. Besides, there is a special emission mechanism in the cusp region of the Earth's magnetosphere, the bremsstrahlung, also called the braking radiation, which happens when energetic electrons brake in an increasing magnetic field. Specific radiation properties in planetary space environment will be introduced in the following sections.

3.3 Radiations of planetary atmosphere

Planetary atmosphere, which originates from the planet, is controlled in different degrees by external sources (e.g., solar activity) and internal sources (e.g., surface and interior) and exhibits diverse complicated variations. Planetary atmosphere is composed of different gases of atomic, molecular, and ionic states, some are active under the effect of sun light and thermal radiation, some are noble compounds (e.g., noble gas), some are chemical active (react with other chemical compositions), and some are compressible gases. The spatiotemporal variations of the atmospheric composition depend on a variety of factors, and this determines the distribution of energy sources within the atmosphere, for example, deposition of solar radiations induce vertical heating, infrared thermal radiations induce cooling, albedo difference due to land or sea results in horizontal distribution.

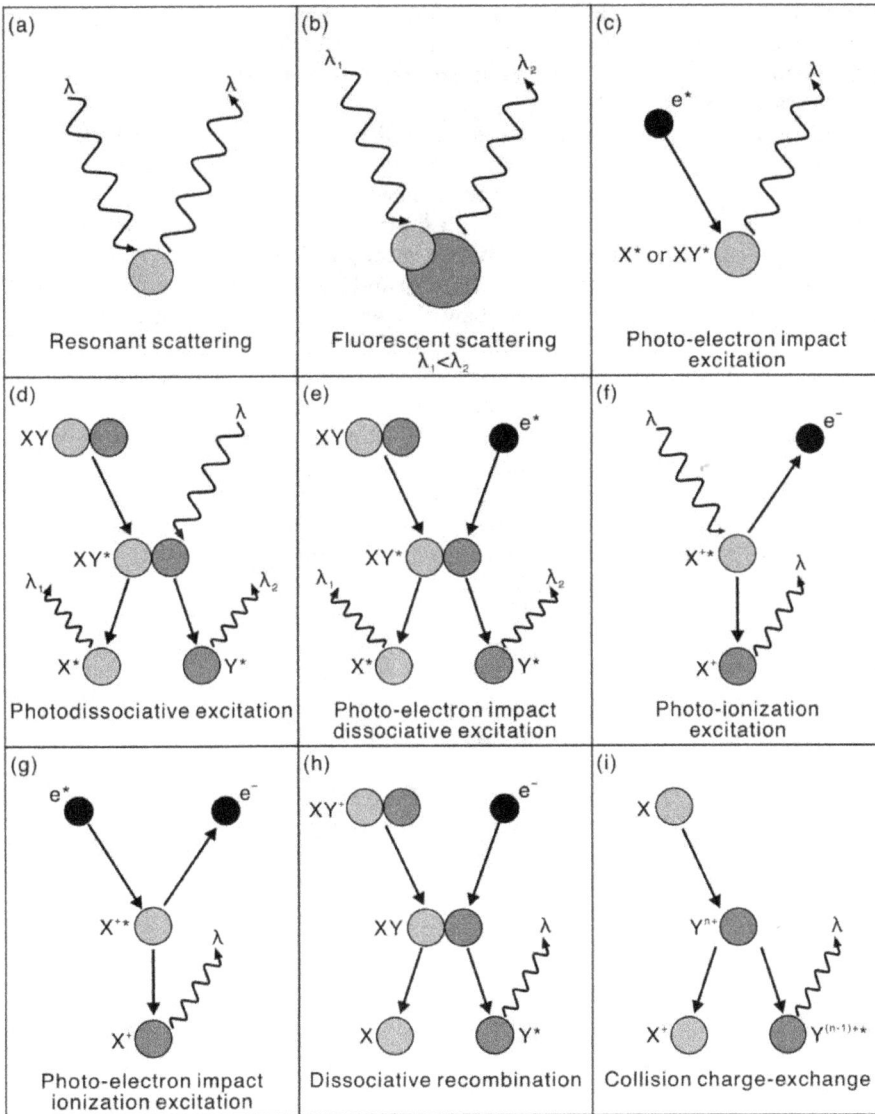

Figure 3.
Typical emission mechanisms in space environment. λ, λ_1, and λ_2 are the wavelength of the incident or emitted photons. X and Y are atoms; XY is a molecule; X^+, Y^{n+}, or $Y^{(n-1)+}$ is an positive ion; and e- and e^* are electrons and photon electrons, respectively. X^*, Y^*, XY^*, X^{+*}, and $Y^{(n-1)+*}$ is an atom/molecule/ion in an excited state. Modified from reference [4]. From a to i, shown are emission mechanisms of resonant scattering, fluorescent scattering, photo-electron impact excitation, photodissociative excitation, photo-electron impact dissociative excitation, photo-ionization excitation, photo-electron impact ionization excitation, dissociative recombination, and collision charge-exchange excitation, respectively.

Based on ground-based telescopes, circling or flyby spacecraft, using spectrometric and occultation technologies (solar occultation, star occultation or spacecraft occultation), the atmospheric composition can be determined. In combination of *in-situ* mass spectrometer, the composition of planetary atmosphere can be accurately determined (even the isotopic level). When sufficient observation samples are obtained, the spatiotemporal distribution of the active compounds in the atmosphere can be determined. Using more advanced spectrographic imaging technology, the atmospheric compositions and their spatiotemporal variations can be obtained simultaneously to investigate the atmospheric dynamics. The basis for optical remote sensing is the continuous spectrum and discrete spectrum of the planetary atmosphere.

The continuum spectrum of a planet, from X-rays to radio wavelength, originates from a combination of its surface, atmosphere, and space plasma. The continuum spectrum is composed of two parts. The first part is the reflected spectrum of the sunlight radiation (or from star for exoplanets) by the gases and particles in the atmosphere and by the surface with typical wavelengths ranging from UV to near infrared (0.4–10.0 μm), where the Sun and stars have their emission peak according to the Wien's Displacement Law. The second part is the thermal radiation spectrum originated from the radiative flux emitted by the planet with typical wavelengths ranging from infrared to microwave (>1–10 μm), depending on the body temperature. The continuum spectrum can be obtained by the summation of a diffuse blackbody mirror of the star spectrum and the blackbody of the thermal spectrum. Typical spectrum of a planet is shown in **Figure 4**. The spectrum for various planets are significantly different due to the factors such as the distant to the star, the atmospheric composition, and the temperature of the planet. At short wavelengths (mainly UV and visible), excitations of atoms, molecules, or ions in the atmosphere by the absorption of solar photons or the precipitation of particles generate emission spectrum (e.g., fluorescent emission and aurora phenomena). At radiofrequencies, the planetary emission is dominated by nonthermal processes occurring in planetary magnetic field, mainly by electrons gyrating in spirals along magnetic field lines, for example, the synchrotron emission in the case of Jupiter.

The discrete spectrum also contains two parts. On the one hand, emission lines are produced by atoms, molecules, and ions excited by a variety of physical and chemical processes. On the other hand, the absorption lines and bands are produced by atoms and molecules. The emission lines and absorption lines and bands are the 'fingerprint' of atmosphere. Specific lines may indicate the existence of corresponding chemical compounds or certain states of atoms and molecules. Therefore, such 'fingerprint' is usually used to discovery and identify the specific composition

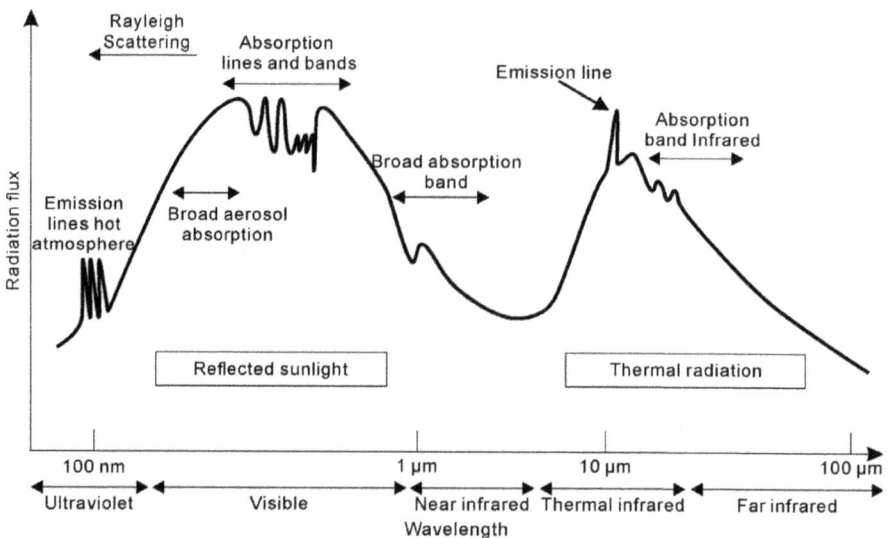

Figure 4.
Typical spectrum of a planet with atmosphere [7].

and its amount in an atmosphere. For example, the discovery of H_2O and CH_4 in Martian atmosphere.

Specifically, the spatiotemporal distribution of certain tracer composition can be obtained through imaging at corresponding narrow band emission lines to investigate planetary atmospheric dynamics. For example, the airglow emissions of atomic oxygen at 557.7 nm and 630.0 nm are widely used to investigate the dynamics of thermosphere and ionosphere and the auroral physics.

3.4 Radiations of planetary space plasmas

The X-ray and UV radiations from the Sun or the stars ionize the upper atmosphere of a planet, the ions and electrons form a quasi-stable plasma distribution in planetary space (e.g., ionosphere, plasmasphere, radiation belt, and plasma sheet in the case of Earth) under the combined effects of solar wind, planetary rotation, intrinsic magnetic field, and other factors. Due to the different physical properties of planets, especially the different configurations of intrinsic magnetic field and the different distances to the Sun or the stars, the plasma composition, energy, distribution, and activity exhibit large distinction (**Table 1**). The most significant manifestation of such distinction is that, under the drives of solar wind disturbances and/or internal sources of planetary system, the plasmas exhibit global or local flow, acceleration, and loss, and thus resulting in mass transportation, energy deposition/dissipation processes with different temporal/spatial scales. The strong interaction between the plasmas trapped by planetary magnetic field and the atmosphere leads to the heating of upper atmosphere, generation of neutral wind, ionization of neutral gasses. Energetic ions and electrons that precipitate into the atmosphere remarkably modify the atmospheric chemistry. During the evolution process of a planet, interaction between plasma and neutral atmosphere significantly contributes to the isotopic fractionation. Bombardment of energetic particles on the surface of planet and its moons will obviously modify the surface property and change the albedo and spectrum characteristics. Traditional *in-situ* field and particle detections are difficult to capture the global view of the mass and energy transportation and are also hard to separate the temporal and spatial variations of the space plasmas, thus limited our understanding of the global coupling dynamics. Optical remote sensing is an important method to overcome these difficulties.

In geospace, in most of the magnetospheric region from the ionosphere to the magnetosheath, plasmas of different property have their characteristic optical radiation. Optical imaging at different wavelengths can be used to answer different scientific questions. For example:

1. EUV and soft X-ray emissions in the terrestrial magnetosheath and cusp regions. Collision charge exchange between the solar wind He^{2+} and geocoronal atomic hydrogen emits photons at 30.4 nm, and when the same process happens between the highly charged heavy solar wind ions (e.g., O^{7+} and C^{6+}) and geocoronal atomic hydrogen photons in X-ray are emitted [5]. Global imaging of the magnetosheath at 30.4 nm and X-ray can visualize the three-dimensional structure of the bow shock and magnetopause, reveal the dynamical process of the entry of solar wind mass and energy into the magnetosphere through the solar wind-magnetosphere interaction, and provide accurate input for the prediction of near-Earth space weather.

2. EUV emissions in the magnetosphere. The plasmaspheric He^+ and the magnetospheric O^+ resonantly scatter the sunlight at 30.4 nm 83.4 nm,

respectively [3, 5, 8]. Global imaging of the plasmasphere and the magneto-sphere at 30.4 nm and 83.4 nm can visualize the large-scale convection in the magnetosphere and reveal the dynamic variation terrestrial matter.

3. Auroral emissions in polar region. The auroral oval is a projection of solar wind/magnetosphere energetic particles along the geomagnetic field lines. The precipitated energetic particles collide with atoms, molecules and ions in the upper atmosphere and generate auroral emissions in wavelengths from X-ray to near infrared (the X-ray is generated by energetic electrons through brems-strahlung) [8]. Multi-wavelength global imaging of the aurora oval can be used to establish the relationship between auroral activity and space mass/energy transportation, break through the limitations of *in-situ* measurements.

In the space environment of other planets, optical remote sensing can also be used to obtain the global image. For example:

1. The solar wind particles bombard the Mercury surface to sputter out sodium atoms. The following interaction with the solar wind generates a sodium exosphere and a sodium tail downstream of Mercury as long as 1,400 Mercury radii. Global images at the sodium doublet (589.0 nm and 589.6 nm) [9] can be used to investigate the solar wind sputtering process and the global dynamics sodium exosphere and tail.

2. Under the effects of solar radiations and energetic particles, photochemical, collision, and other processes excite EUV-UV–Visible emissions in Venusian at-mosphere and ionosphere [10], which can be used to investigate the evolution characteristics of Venusian space environment and the interaction between solar wind and planet without intrinsic magnetic field but dense atmosphere.

3. In the case of Mars, the charge exchange collision between solar wind ener-getic protons and neutral hydrogens emits photons at 121.6 nm, other emission mechanisms illustrated in **Figure 3** can generate O^+ 135.6 nm line, CO+ 2 UV emission bands, O 557.7 nm line, and other bands [6, 11], which can be used to investigate the interaction between solar wind and planet with local crustal magnetic field and thin atmosphere and the global view of Martian atmo-spheric escape.

4. The mass released from the volcanic activity of Io, the first moon of Jupiter, forms a plasma torus at ~7 R_J (Jupiter radii) in Jovian magnetosphere. The Io plasma torus controls the dynamics of Jupiter's magnetosphere and only opti-cal imaging can capture its global evolution picture. Coordinated with X-ray and far UV imaging of Jupiter aurora, we can investigate how Io's volcanic activity affects the magnetospheric plasma source, as well as the subsequent evolutions. The operational wavelengths include 673.1 nm (S^+), 68.5 nm and 953.1 nm (S^{2+}) in Io plasma torus, sodium doublet in Jupiter's neutral nebula, and the X-ray to far UV auroral emissions [12–15].

5. Like the Jupiter system, the space environment of Saturn system is also affect-ed by geological activities of its moons. The rings and the water erupted from Enceladus are the main plasma sources in Saturn's magnetosphere. Optical imaging of the O^+, H_2O^+, and other water-based ions at EUV-UV wavelengths can be used to investigate how Enceladus's geological activity affects the magnetospheric plasma source, as well as the subsequent evolutions.

3.5 Radiations of planetary geological activities

Geological activity of planet and its moons plays a key role in the evolution of the planetary system. Planetary geology research helps to understand the formation and evolution of celestial bodies in the solar system, deeply understand the evolution of the Earth, and reveal the origin and evolution of life on the Earth [16]. Compared with traditional geological techniques, such as sample analysis, analogy study, and simulation, remote sensing is of great importance to the acquisition of matter composition, structural characteristics, geological history and so on.

Measurements of the planetary thermal radiation intensity in infrared can be used to obtain the temperature and composition of planetary atmosphere and surface, the thermal moment, and the property and evolution of planetary surface. The reflected spectrum in visible to infrared range can be used to obtain the chemistry, mineralogy (silicate) and regolith maturity of surface, the surface geology, and the degree of planetary differentiation. Imaging and spectrographic imaging at UV, visible and infrared can be used to obtain surface properties, relative age, surface action, and history of the planet. With a laser altimeter or a synthetic aperture radar, the surface relief can be acquired. In combination with gravitation data, the isostatic action can be inferred. With a UV–visible photometer, the matter property of planetary surface can be obtained, and the surface composition and differentiation can be induced. The X-ray and gamma ray spectra can be used to measure the abundance of K, U, Th, and other elements that are bombarded by cosmic rays to induce the surface composition, property, thermal history, and differentiation of a planet. Finally, the combination of spectrum and imaging of different wavebands can provide a more detailed study of the geology, origin, and evolution history of planets.

3.6 Optical signals of life

As an important research field in planetary science, the astrobiology mainly focuses on the origin and evolution of life on Earth and other solar system bodies and the exploration of potential distribution and future trend of life in the universe. This field involves astronomy, geology, life science, and other disciplines, and the subjects include the origin and early evolution of life on Earth, search and study of habitable planets (environment) and potential life forms, co-evolution of life and environment, and artificial construction of habitable environment [16–18]. Search for signs of life on planet always lies in the core of planetary science. The most direct way, of course, is collection of life samples. However, no life sample was found on the extraterrestrial celestial bodies that human spacecraft have visited. Discoveries of water and methane on Mars have greatly stimulated the desire of search for life on Mars in the past decades. The paleolakes or paleo-oceans on the Mars and liquid oceans on Europa and Enceladus have pushed the search for extraterrestrial life to the core of planetary exploration.

Beyond the solar system, in places where human spacecraft cannot currently reach, such as exoplanets, the only reliable way to search for signs of life is remote sensing. For thousands of years, human beings have been trying to figure out whether they are the only intelligent beings alone in the universe. The 2019 Nobel Prize in physics awarded M. Mayor and D. Queloz who discovered a planet outside our solar system, known as an exoplanet, around a sun-like star in October 1995. This discovery is a firm step for the exploration of life on exoplanets, and it opens the door to the detection of life on exoplanets. By the end of 2020, 4,374 exoplanets have been discovered from 3,234 planetary systems. Exoplanet life can only be detected by searching for signs of life, such as organic molecules or other indicators of the presence and activity of lives, on the planet's surface or in the atmosphere. The detection of exoplanet atmosphere mainly relies on optical, UV and near-infrared bands, among which molecules related to life activities and relatively easy to be detected include O_2, O_3, H_2O, CO_2, CH_4, NO_2 and so on. Of course, even these signals are detected, other information such as the nature of the planet itself, the nature of the star, and the properties of planetary systems are needed to make a comprehensive decision. Currently, acquisition of these information mainly relies on optical methods, such as radial velocity, microlensing, transit, and imaging [19].

The radial velocity method measures the change of the radial velocity of the host star caused by the gravitational force of the planet (a Doppler effect of light) to estimate the mass of the planet. Until the Kepler space telescope was launched, it was the most effective way to identify exoplanets, including Pegasus 51b, the first exoplanet orbiting a sun-like star.

The microlensing method measures the bending and amplification of light in gravitational fields to detect objects, including exoplanets. Gravitational lensing is an optical effect predicted by Einstein's general theory of relativity. Since the foreground object passing through the background star is accidental, the application of microlensing method is also accidental and unrepeatable, which has a great impact on the accuracy of exoplanet detection. This method is particularly sensitive to the detection of cold planets (i.e., large orbital radii).

The transit method detects exoplanets and determines their size by measuring the periodically dimming brightness of a star due to the obscuration of the exoplanet in line of sight. By measuring the tiny changes in the brightness of the target star, we can detect telltale signs of an exoplanet in the light curve. The first exoplanet discovered by this method is HD 209458b, a hot Jupiter discovered in 1999. With the launch of the Kepler space telescope, transit method has become the most prolix method of finding exoplanets. By the end of 2020, more than 3,100 exoplanets have been discovered through transit method.

Direct Imaging, as the name suggests, is the direct optical imaging of exoplanets. For general main-sequence stars, the thermal radiation is mainly concentrated in the UV to near-infrared band and the peak value is between visible and UV band according to Stepan-Boltzmann's Law and Wien's Displacement Law. Exoplanets, however, do not have sufficient and stable energy sources, generally have low temperatures, and their thermal radiation is mainly concentrated in the infrared band. So, in the case of exoplanet with large radiation fluxes, we can distinguish two exoplanets by looking at infrared wavelength. On the other hand, the observation requires high performance instrument, a coronograph to block the light from the stars, and the observation system needs to be maintained at a very low temperature to reduce the infrared radiation from the instrument. In general, the direct imaging method is used to search for young Jovian planets with temperatures between 600 and 2000 K, and the peak wavelength of thermal radiation is between 1.4 and 4.8 μm. Such planets are usually far enough from their parent star with large enough surface areas and radiation fluxes and can be observed in the near-infrared to mid-infrared wavelengths.

The first exoplanet was discovered using the radial velocity method, but the method that discovered the largest number of exoplanets is transit. It is expected that transit will continue to be the most effective observation method in the next 15 years. In order to further study the nature of the atmospheres of exoplanets, especially terrestrial exoplanets, both transit and direct imaging are necessary, and the latter will be the most promising observation method in the future.

4. Strategy and plan in China

From the development of planetary science and planetary exploration for more than 400 years, no matter it is space-based or ground-based, optical remote sensing has always occupied an important position. With the development of optical, mechanical and electronic technologies, the aperture of ground-based optical remote sensing equipment has developed from a few centimeters to tens of meters, and its resolution and sensitivity have been improved by orders of magnitude. Advanced remote sensing and *in-situ* detectors carried by various satellites have made our understanding of the planetary environment reach an unprecedented height. However, it is not hard to see that in this development process, especially in the six decades since the Space Age, there has been little presence or voice of China. With the deep space exploration becoming a national strategy, planetary exploration and planetary science will become the hot spot of scientific and technological development in the future.

On January 6, 2019, the Degree Evaluation Committee of University of Chinese Academy of Sciences (UCAS) approved the plan to set up the planetary science as a first-level discipline, as proposed by the Institute of Geology and Geophysics, Chinese Academy of Sciences (IGGCAS). This plan has also been submitted to the State Council Academic Degrees Committee, who will finally approve the establishment of the new first-level discipline in China. This marks a new period of historical opportunity for the development of planetary science in China. On July 2, 2019, sponsored by the College of Earth and Planetary Science, UCAS, colleges of 27 universities established a China University Planetary Science Alliance in Beijing, targeted at establishing and improving the layout of planetary science discipline in China, perfecting the talent training and scientific research system of planetary science, promoting the coordinated development of planetary science and exploration technology. Taking this as an opportunity and relying on China's intensive deep space exploration missions to carry out ground-based, aero-based and space-based optical remote sensing of planetary space environment, it will greatly promote the integrated development of science and education in planetary science in China [20–23].

4.1 Lenghu observatory for planetary science

In order to support the science and education integration strategy of planetary science in China, it is urgent to build planetary observation facilities. Under the current situation, it is of practical significance to develop ground-based optical remote sensing for planetary science to quickly gather planetary research teams and train planetary exploration and scientific research personnel. At present, almost all optical telescopes in the world are built by astronomers, and there is no ground-based telescope dedicated to planetary science. It is imperative to build dedicated planetary telescopes in China. In China's planetary exploration roadmap, ground-based optical observation is also clearly taken as the first step [20]. To support the development of planetary science exploration and research, the IGGCAS established the Lenghu Observatory for Planetary Science (IGGCAS-LOPS) in the Lenghu Astronomical Observation Base (LAOB) in 2020. Based on the high-quality site conditions of LAOB on Saishiteng Mountain, Lenghu Town, Qinghai Province [24], the IGGCAS-LOPS will build planetary optical telescopes to carry out ground-based observation. The IGGCAS-LOPS currently has two meter-sized telescopes, the 0.8 m Planetary Atmosphere Spectroscopic Telescope (PAST) and the 1.8 m Io geological activity observatory (TINTIN).

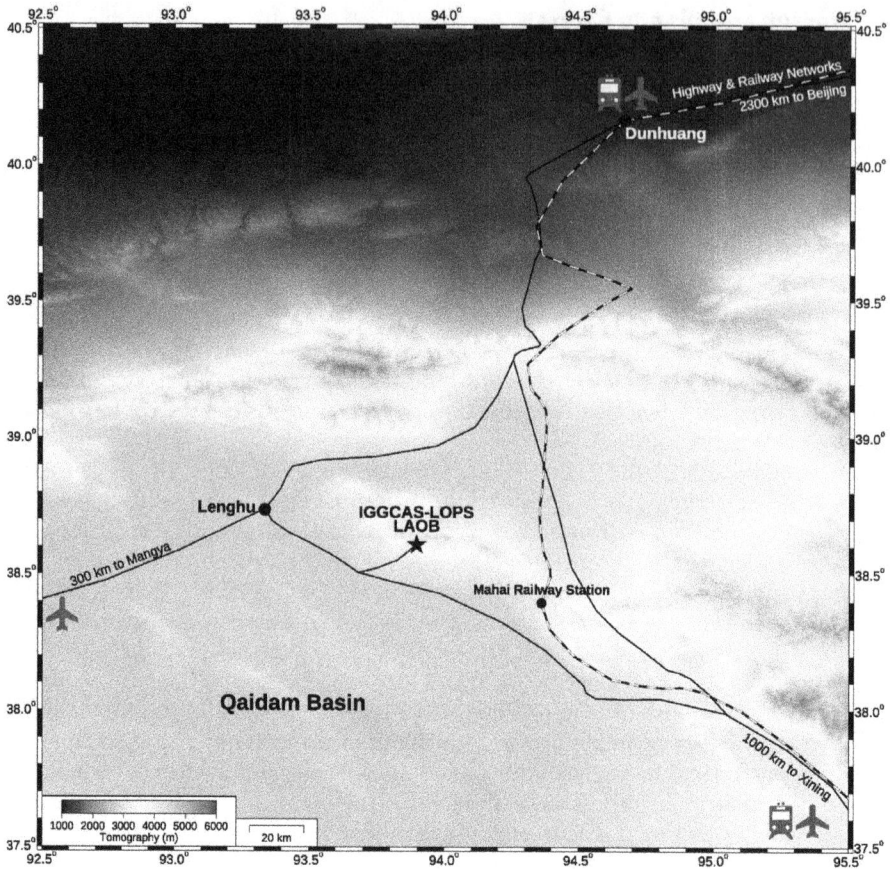

Figure 5.
Basic geographic information for the IGGCAS-LOPS in the LOAB (modified from reference [24]).

The IGGCAS-LOPS is located on the 4,200 m altitude platform of Saishiteng Mountain, about 80 km east of Lenghu Town (as shown in **Figure 5**). The environmental monitoring data since 2018 show that the platform has excellent atmospheric visibility and very good nighttime seeing [24], and it is one of the few sites in the world that can be used for optical imaging of planetary space environment. The site is also an astronomical observation base built by Qinghai Province with great efforts. Astronomy and planetary science will become the two core business cards of the base.

4.2 Telescopes at IGGCAS-LOPS

The two telescopes currently planned for the IGGCAS-LOPS are PAST and TINTIN. The PAST (shown in **Figure 6**) was a 0.8 m UV–visible telescope built with the support of the strategic priority research program of CAS, the near space science experiment system, and was developed by the Changchun Institute of Optics, Fine Mechanics and Physics, Chinese Academy of Sciences. The main scientific objectives of the telescope are the orbital motion characteristics, atmospheric and plasma distribution characteristics, and spectral radiation characteristics of the celestial bodies within the orbit of Jupiter in the solar system. The telescope is planned to operate on both a balloon platform and a ground station. Through

Figure 6.
Illustration of PAST.

comprehensive consideration, the key parameters of the system are determined. The aperture of the telescope is 0.8 m, the operating waveband is 280–680 nm (multiple narrowband filters), the field of view is 15′, the angular resolution is 0.5″. The telescope was also equipped with a Jupiter coronagraph, which will be able to greatly attenuate the intense radiation from Jupiter itself and observe the faint atmosphere and plasma radiation around the planet. The PAST is scheduled to be installed at the station in June 2021. Upon installation, observations of the Jupiter system will begin immediately. At the same time, it can also carry out observations of comets and small celestial bodies.

Another telescope, TINTIN, will be purchased from Germany company Astelco. The aperture of TINTIN is 1.8 m, with a spectral range of 392–1100 nm (as shown in **Figure 7**). The telescope is equipped with a scientific camera, a Jupiter coronagraph, and an echelle spectrograph. The telescope's core goals are to monitor Io's atmospheric escape from volcanic activity and the evolution of Io's plasma torus. The telescope will also be equipped with a coronagraph with a pixel resolution of 0.25″ and a FOV of 5′. The multi-scale monitoring of the evolution of mass and energy from Io's geological activity to plasma torus in Jupiter's space will be performed for the first time. The observations can be used to investigate how a moon's geological activity couples with the planetary space environment, in combination with current/future international/China's Jupiter exploration programs. The construction of the observation tower and dome will begin in summer of 2021. The overall construction of TINTIN project will be completed by the end of 2022, when scheduled observations will also begin.

The routine observations of the two telescopes will mainly be realized by remote control, which is locate at IGGCAS, Beijing. At the same time, we will also set offices in the Lenghu Town with the help of local government to support the regular and irregular operation and maintenance of the telescopes. In response to the large amount of data flow of the telescopes, two data centers will be established at IGGCAS and Lenghu Town, respectively. Both data centers will be connected to the server of the telescopes via high-speed networks to ensure that the observations can be timely delivered to the scientific teams to achieve timely scientific impact.

Figure 7.
Illustration of TINTIN.

5. Optical remote sensing images of planetary space

In this section, optical remote sensing images of planetary space environment will be briefly introduced to demonstrate different applications on different planets. Not all the planets in the solar system will be introduced here, only the Earth and Jupiter are taken as examples. For other planets, one can refer to Section 3 for detailed information.

Imaging of the Earth's ionosphere, aurora, and plasmasphere have been developed for decades. **Figure 8a-d** shows examples of the most recent ionospheric image from the Global-scale Observations of the Limb and Disk UV spectrograph (GOLD UVS), the auroral image in far UV from the wide-angle auroral imager onboard the Chinese Fengyun-3D satellite (Fengyun-3D WAI), and the plasmaspheric images at 30.4 nm from the Extreme Ultraviolet Imager onboard the Inner Magnetosphere-Aurora Global Explore (IMAGE EUVI) and the Moon-based Extreme UV Camera onboard the Chinese Chang'e-3 lunar lander (Chang'e-3 EUVC), respectively. Principles of these optical emissions have been introduced in Section 3.4. The nighttime ionospheric disk image can be used to investigate the structure and

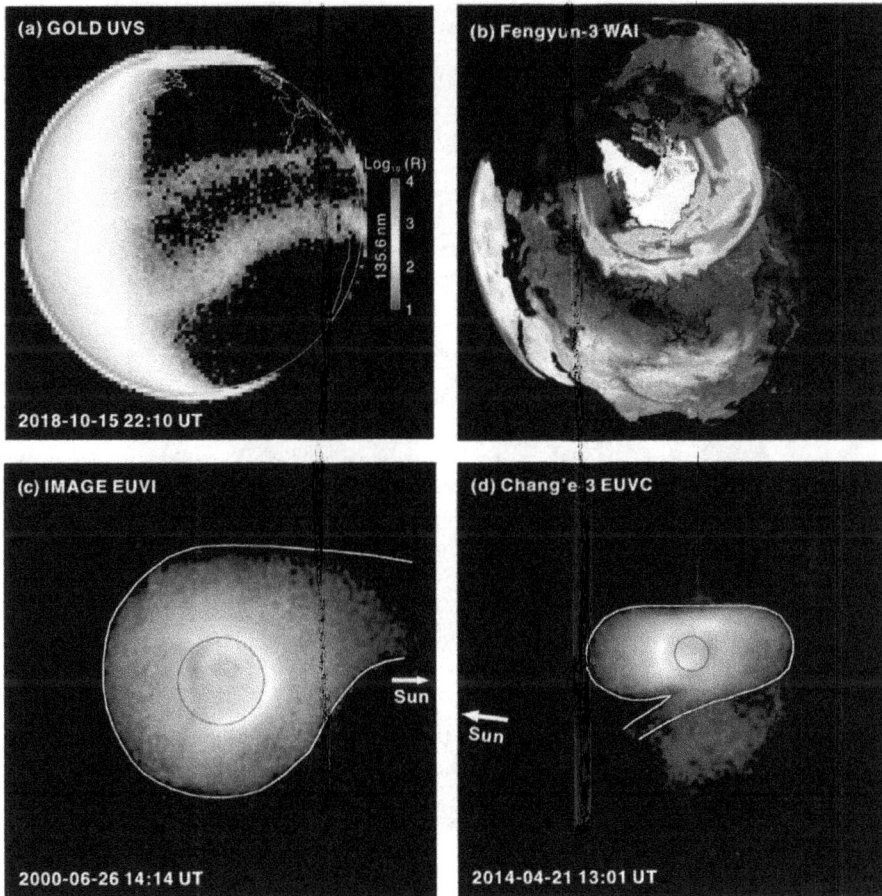

Figure 8.
Optical remote sensing images of Earth's space environment. Shown are (a) nighttime ionospheric disk image at OI 135.6 nm obtained by GOLD UVS (adapted from reference [25]), (b) auroral disk image at N_2 LBH band obtained by Fengyun-3D WAI, and plasmaspheric images at 30.4 nm obtained by (c) the IMAGE EUVI and (d) the Chang'e-3 EUVC.

evolution of equatorial ionization anomaly [25]. The auroral image can be used to manifest the thermal plasma transportation in the Earth's magnetosphere and to denote the substorm activity in Earth's space [26]. The plasmaspheric images can be used to characterize the large-scale convections in the Earth's magnetosphere [27, 28]. When the Chinese Academy of Science (CAS) and European Space Agency (ESA) collaborative science mission – Solar wind Magnetosphere Ionosphere Link Explorer (SMILE) – is launched in the following years, X-ray imaging of the magnetosheath and polar cusp region will be realized. Until then, comprehensive investigations with all the optical remote sensing images will significantly enhance our understandings on the global dynamics of the Earth's magnetosphere.

Figure 9 shows examples of optical remote sensing images in Jovian magnetosphere. The major plasma source in the Jovian magnetosphere is the Io plasma torus, while the major driver for fundamental plasma processes is planetary rotation, very different from the case in terrestrial magnetosphere. Neutral gases released during the volcanic eruptions on Io escape into Jupiter's space. These gases are then ionized by solar radiations. Finally, the ions are trapped by the strong magnetic field of Jupiter and form a plasma torus due to the fast planetary rotation, as shown at the bottom of

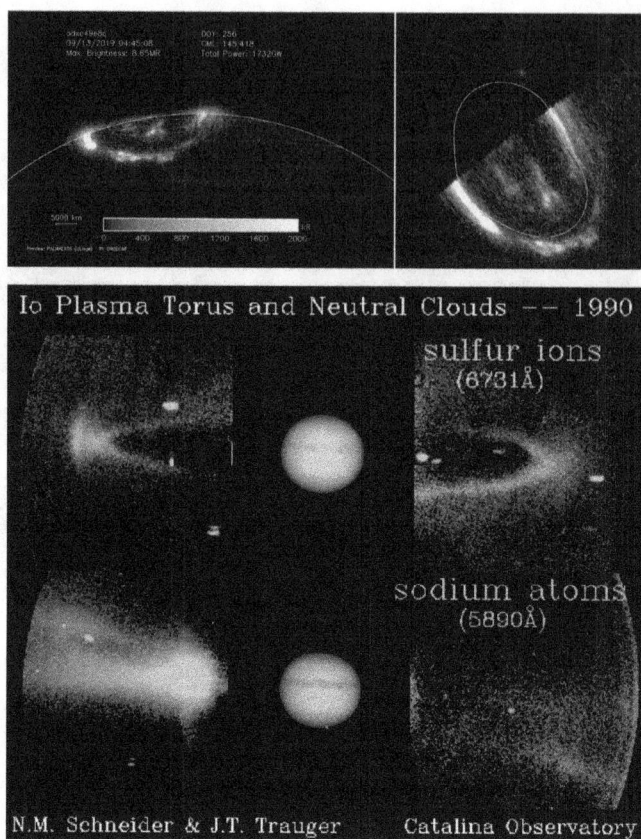

Figure 9.
Optical remote sensing images of Jupiter space environment. The top panel shows UV auroral images captured by the Hubble space telescope (modified from reference [29]). (left) the view from earth orbit and (right) a projection to the northern polar region. The middle and bottom panels show images of the global structure of the Io plasma torus at S⁺ 673.1 nm and Na 589.0 nm (credit: University of Colorado/Catalina Observatory/N. Schneider).

Figure 9, in which the Io plasma torus is imaged at 673.1 nm emitted by singly ionized sulfur. The planetary rotation-driven plasma processes then transport the plasmas into the polar region along dipolar magnetic field lines to form the most powerful aurorae in the solar system, as shown at the top of **Figure 9**, in which the UV aurora is observed by the Hubble Space Telescope (HST) [29]. Coordinated observations of the Io plasma torus (e.g., the two telescopes at LOPS) and the Jovian aurorae (e.g., the HST or the Juno spacecraft) in different temporal and spatial scales will help reveal the mass and energy transportation patterns in Jovian space environment.

6. Summary and outlook

As we all know, forward-looking and feasible scientific objectives are the core factors to ensure that a research project can achieve high scientific output. Currently, the IGGCAS-LOPS is preparing a scientific team to set scientific objectives for future observations. In order to ensure that the scientific team is cutting-edge and experienced, the team will be composed of well-known domestic and international planetary scientists, whose research fields cover planetary space physics, planetary geology, astronomical observation and optical remote sensing technology. Through regular scientific team meetings, the telescope's observation schedules are adjusted in time, with particular attention paid to observation objects with scientific timeliness and social impact. In addition, the international team will organize special sessions at major international conferences, such as the annual meetings of the American Geophysical Union and the European Geosciences Union, to expand the international influence of the IGGCAS-LOPS.

The PAST is expected to begin regular observations in June 2021, and the TINTIN will begin regular observations in 2023. The long-term coordinated observations with PAST and TINTIN will ensure the investigation of the evolution of Io plasma torus in Jupiter's magnetosphere at different temporal and spatial scales. Apart from the primary object of Jupiter system, the secondary object is Mars. The airglow of Martian atmosphere/ionosphere in near UV and visible will be imaged according to Mars phase. Both telescopes can also be used to monitor comets, asteroids, and other celestial bodies in the solar system.

In the future, we will further upgrade the performance of the telescopes and add new instruments to expand the observable targets of the telescope. Compared with other astronomical survey telescopes at Lenghu, the PAST and TINTIN can be considered as precision photometric telescopes, which can cooperate with other astronomical survey telescopes at Lenghu to track and observe their survey targets. The diameter of the two telescopes is also large enough to observe exoplanets, and the radial velocity and transit methods can be used to carry out the search and identification of exoplanets.

In addition to providing first-hand data from autonomic planetary observations for related scientific research, the IGGCAS-LOPS is also an important scientific and educational practice base. Considering that the Qaidam Basin near Lenghu is the largest Mars-like geomorphologic environment in the world, the IGGCAS-LOPS and its surrounding geographical environment will become a comprehensive practice site for planetary geology and planetary space science. In July 2020, the College of Earth and Planetary Sciences, UCAS and the government of Haixi Mongolia and Tibetan Autonomous Region of Qinghai Province have signed an agreement on a planetary science practice base. Upon the completion of the telescopes, the IGGCAS-LOPS will provide important scientific practice support. In addition, the IGGCAS-LOPS is also considering building planetary practice bases with other universities.

In the long run, there is still a lack of comprehensive earth and planetary space environment observation station in the vast western region of China. Based on the IGGCAS-LOPS, further deployment of atmospheric, ionospheric, geomagnetic, seismic, and other observation equipment is planned. The excellent atmospheric optical conditions and extremely low light pollution at LAOB are very suitable for the monitoring of upper atmospheric glow in the mid-latitude region and the study of the dynamics of the middle and upper atmosphere. At the same time, together with ionospheric vertical detection (altimeter, radar, etc.) and geomagnetic field detectors, we can comprehensively study the dynamics of atmospheric and ionospheric vertical coupling and cooperate with a series of stations in eastern China to form a more complete coverage of China's space environment.

Acknowledgements

This work is supported by the Key Research Program of the Institute of Geology & Geophysics, CAS (grant IGGCAS-201904). All the WAI raw data were processed and provided by the ground application system at NSMC, CMA. The authors sincerely thank the National Astronomical Observatories, Chinese Academy of Sciences, for provision of the CE 3 EUVC data; T. Forrester of IMAGE EUV team for provision of the IMAGE EUV data and relevant processing software.

Conflict of interest

The author declares no conflict of interest.

Author details

Fei He*, Zhonghua Yao and Yong Wei
Institute of Geology and Geophysics, Chinese Academy of Sciences, Beijing, China

*Address all correspondence to: hefei@mail.iggcas.ac.cn

IntechOpen

Optical Remote Sensing of Planetary Space Environment
DOI: http://dx.doi.org/10.5772/intechopen.98427

References

[1] Wu FY, Wei Y, Song YH, et al. From fusion of research and teaching to leading of science: Strategy to build planetary science program with Chinese characteristics. Bulletin of Chinese Academy of Sciences. 2019; 34(7): 741-747 (in Chinese). DOI: 10.16418/j.issn.1000-3045.2019.07.002

[2] Kivelson MG, Bagenal F. Planetary magnetosphere. In: McFadden LA, Weissman PR, Johnson TV, editors. Encyclopedia of the Solar System. New York: Academic Press, 2007. p. 519-540. DOI: 10.1016/B978-0-12-415845-0.00007-4

[3] He F, Zhang X X, Chen B, Fok MC, Zou YL. Moon-based EUV imaging of the Earth's plasmasphere: Model simulations. Journal of Geophysical Research: Space Physics, 2011; 118:7085-7103. DOI: 10.1002/2013JA018962

[4] McClintock WE, Schneider NM, Holsclaw GM, Clarke JT, Hoskins AC, et al. The Imaging Ultraviolet Spectrograph (IUVS) for the MAVEN mission. Space Science Reviews, 2015; 195:75-124. DOI: 10.1007/s11214-014-0098-7

[5] He F, Zhang XX, Wang XY, Chen B. EUV emissions from solar wind charge exchange in the Earth's magnetosheath: Three-dimensional global hybrid simulation. Journal of Geophysical Research: Space Physics, 2015; 120: 138-156. DOI: 10.1002/2014JA020521

[6] Deighan JI, Jain SK, Chaffin MS, Fang X, Halekas JS, Clarke JT, et al. Discovery of a proton aurora at Mars. Nature Astronomy, 2018; 2:802-807. DOI: 10.1038/s41550-018-0538-5

[7] Sánchez-Lavega A. An introduction to planetary atmosphere. Boca Raton: Taylor & Francis. 2011. p. 115-164. DOI: 10.1201/9781439894668

[8] Meier RR. Ultraviolet spectroscopy and remote sensing of the upper atmosphere. Space Science Reviews, 1991; 58:1-158. DOI: 10.1007/BF01206000

[9] Baumgardner J, Wilson J, Mendillo M. Imaging the sources and full extent of the sodium tail of the planet Mercury. Geophysical Research Letters, 2008; 35:L03201. DOI: 10.1029/2007GL032337

[10] Nara Y, Yoshikawa I, Yoshioka K, et al. Extreme ultraviolet spectra of Venusian airglow observed by EXCEED. Icarus, 2018, 307:207-215. DOI: 10.1016/j.icarus.2017.10.028

[11] Gérard JC, Aoki S, Willame Y, et al. Detection of green line emission in the dayside atmosphere of Mars from NOMAD-TGO observations. Nature Astronomy, 2020; 4:1049-1052. DOI: 10.1038/s41550-020-1123-2

[12] Schneider NM, Trauger JT, Wilson JK, et al. Molecular origin of Io's fast sodium. Science, 1991; 253:1394-1397. DOI: 10.1126/science.253.5026.1394

[13] Mendillo M, Flynn B, Baumgardner J. Imaging observations of Jupiter's sodium magneto-nebula during the Ulysses encounter. Science, 1992. 257:1510-1512. DOI: 10.1126/science.257.5076.1510

[14] Gladstone GR, Waite Jr JH, Grodent D, et al. A pulsating auroral X-ray hot spot on Jupiter. Nature, 2002; 415:1000-1003. DOI: 10.1038/4151000a

[15] Grodent D, Bonfond B, Yao Z, et al. Jupiter's aurora observed with HST during Juno orbits 3 to 7. Journal of Geophysical Research: Space Physics, 2018; 123(5):3299-3319. DOI: 10.1002/2017JA025046

[16] Li X, Lin W, Xiao Z, et al. Planetary geology: "Extraterrestrial" mode of

geology. Bulletin of Chinese Academy of Sciences, 2019; 34(7):776-784 (in Chinese). DOI: 10.16418/j. issn.1000-3045.2019.07.007

[17] Lin W. Life in the near space and implications for astrobiology. Chinese Science Bulletin, 2020; 65(14):1297-1304 (in Chinese). DOI: 10.1360/TB-2019-0805

[18] Lin W, Li Y, Wang G, et al. Overview and perspective of astrobiology. Chinese Science Bulletin, 2020; 65(5):380-391 (in Chinese). DOI: 10.1360/TB-2019-0396

[19] Doyle LR, Deeg HJ. The way to circumbinary planets. In: Deeg HJ, Belmonte JA, editors. Handbook of Exoplanets. Switzerland: Springer, 2018. p. 65-84. DOI: 10.1007/978-3-319-55333-7

[20] Wei Y, Yao Z, Wan W. China's roadmap for planetary exploration. Nature Astronomy, 2018; 2:346-348. DOI: 10.1038/s41550-018-0456-6

[21] Rong Z J, Cui J, He F, et al. Status and prospect for Chinese planetary physics. Bulletin of Chinese Academy of Sciences, 2019; 34(7):760-768 (in Chinese). DOI: 10.16418/j. issn.1000-3045.2019.07.005

[22] Wan W, Wei Y, Guo Z, et al. Toward a power of planetary science from a giant of deep space exploration. Bulletin of Chinese Academy of Sciences, 2019; 34(7):748-755 (in Chinese). DOI: 10.16418/j.issn.1000-3045.2019.07.003

[23] Wei Y, Zhu R. Planetary science: Frontier of science and national strategy. Bulletin of Chinese Academy of Sciences, 2019; 34(7):756-759 (in Chinese). DOI: 10.16418/j. issn.1000-3045.2019.07.004

[24] Deng L, Yang F, Chen X, He F, Liu Q, Zhang B, Zhang C, Wang K, Liu N, Ren A, Luo Z, Yan Z, Tian J,

Pan J. Lenghu on the Tibetan Plateau as an astronomical observing site. Nature, 2021; accepted. DOI: xxxx

[25] Eastes RW, Solomon SC, Daniell RE, Anderson DN, Burns AG, England SL, Martinis CR, McClintock WE. Global-scale observations of the equatorial ionization anomaly. Geophysical Research Letters, 2019; 46:9318-9326. DOI: 10.1029/2019GL084199

[26] He F, Guo RL, Dunn WR, Yao ZH, Zhang HS, Hao YX, Shi QQ, Rong ZJ, Liu J, Tian AM, Zhang XX, Wei Y, Zhang YL, Zong QG, Pu ZY, Wan WX. Plasmapause surface wave oscillates the magnetosphere and diffuse aurora. Nature Communications. 2020; 11:1668. DOI: 10.1038/s41467-020-15506-3

[27] Sandel BR, Goldstein J, Gallagher DL, Spasojević M. Extreme ultraviolet imager observations of the structure and dynamics of the plasmasphere. Space Science Reviews, 2003; 109(1):25-46. DOI: 10.1023/B:SPAC.0000007511.47727.5b

[28] He F, Zhang XX, Chen B, Fok MC, Nakano S. Determination of the Earth's plasmapause location from the CE-3 EUVC images. Journal of Geophysical Research: Space Physics, 2016; 121:296-304. DOI:10.1002/2015JA021863

[29] Yao Z H, Bonfond B, Clark G, Grodent D, Dunn WR, Vogt MF, Guo RL, Mauk BH, Connerney JEP, Levin SM, Bolton SJ. Reconnection- and dipolarization-driven auroral dawn storms and injections. Journal of Geophysical Research: Space Physics, 2020; 125:e2019JA027663. DOI: 10.1029/2019JA027663

Chapter 5

Image Enhancement Methods for Remote Sensing: A Survey

Nur Huseyin Kaplan, Isin Erer and Deniz Kumlu

Abstract

The quality of the images obtained from remote sensing devices is very important for many image processing applications. Most of the enhancement methods are based on histogram modification and transform based methods. Histogram modification based methods aim to modify the histogram of the input image to obtain a more uniform distribution. Transform based methods apply a certain transform to the input image and enhance the image in transform domain followed by the inverse transform. In this work, both histogram modification and transform domain methods have been considered, as well as hybrid methods. Moreover, a new hybrid algorithm is proposed for remote sensing image enhancement. Visual comparisons as well as quantitative comparisons have been carried out for different enhancement methods. For objective comparison quality metrics, namely Contrast Gain, Enhancement Measurement, Discrete Entropy and Average Mean Brightness Error have been used. The comparisons show that, the histogram modification methods have a better contrast improvement, while transform domain methods have a better performance in edge enhancement and color preservation. Moreover, hybrid methods which combine the two former approaches have higher potential.

Keywords: Remote Sensing, Image Enhancement, Histogram Modification, Transform Domain Methods, Image Decomposition

1. Introduction

Widely used remote sensing applications, such as mapping, classification, soil moisture detection, target detection and tracking, etc. require high quality images. To meet the increasing need for higher quality images, image enhancement methods which improve the contrast and edge information of the input image are applied to the raw input images.

Images provided by remote sensing devices have to be enhanced by special methods instead of standard enhancement methods. Since applications like classification, target detection and target tracking are automated applications, the original reflectance values of the input image should be preserved as much as possible, which makes enhancing the remotely sensed image a challenging problem [1, 2]. Remote sensing image enhancement techniques should improve the visibility, contrast and edge information of the image while preserving the original reflectance values.

In recent years, many remote sensing image enhancement methods have been developed to increase the quality of these images. Image enhancement methods can be divided into two main groups as direct and indirect methods [3–5]. Direct methods aim to enhance the images by using a defined contrast measure [6–9], while the indirect methods try to improve the dynamic range of the images without a contrast measurement [10–15].

In direct methods, contrast measurements can be global or local. In general, local measurements have better results [9]. Dhnawan et al. [6] proposed a local contrast function based on the relative difference between a central region and a neighboring region for a given pixel. Beghdad and Negrate [7] introduced an improvement of [6] by defining the contrast with the consideration of edge information. Laxmikant Dash and Chatterji [8] proposed an adaptive contrast enhancement method where contrast amplification is based on the brightness estimated by local image statistics. Cheng and Xu [9] proposes a another adaptive enhancement method based on the fuzzy entropy principle and fuzzy set theory.

The direct methods have a low computational cost but accordingly show a poor image enhancement performance. The state of art methods are generally indirect methods which provide better enhancement performances compared to the direct methods. The indirect methods can be divided into two sub categories as histogram modification based methods [3, 4, 16–22] and transform domain methods [1, 2, 21, 23–25].

The simplest histogram modification method is Histogram Equalization (HE) [16]. In this method, the histogram distribution of the input image is aimed to have uniform distribution. This method is able to improve the contrast. However, the HE based enhanced images generally suffer from undersaturation or oversaturation, which results in poor quality images. To fix this problem, more efficient histogram modification methods have been proposed in recent years such as Bi-Histogram equalization (BHE) [17] based and Recursive Mean-Separate histogram equalization (RMSHE) [18]. In both methods, the original histogram of the input image is divided into sub-histograms. After obtaining the sub-histograms, separate histogram equalizations are applied to these sub-histograms. Finally, the divided histograms are merged to obtain the enhanced image [17, 18]. The images obtained by these methods have higher quality compared to the classical HE method, however the undersaturation and oversaturation problems are not resolved. 2-D histogram based methods have also been proposed for image enhancement [19, 20]. These methods provide better results than the methods aforementioned, however the computational cost of 2-D histogram creation is too high, which makes these methods not suitable for automated applications. Moreover, there are faster methods with higher enhancement performances. Another method proposed in this sub-category is Adaptive Gamma Correction with Weighting Distribution (AGCWD) method [4]. In this method, a weighted distribution of the original histogram of the input image is obtained followed by Gamma correction. The most important benefit of this method is its ability to preserve the original reflectance values which are needed for remotely sensed image enhancement, however this method too suffers from saturation artifacts. Moreover, the edge information is lost especially in the brighter regions [2, 21]. Histogram modification methods have a good performance if the histogram of the input image is smoother. Moreover, these group eliminate the lower-scale details [22]. The histogram modification methods have a higher performance for low resolution images and images containing larger-scale details.

Transform domain based image enhancement methods use certain transformations to decompose the image into subbands and improve the contrast by modifying specific components [1, 2, 23–25]. The first method in this category uses a combination of discrete wavelet transform and singular value decomposition (DWT-SVD) [23]. In DWT-SVD method, first discrete wavelet transform (DWT) is applied to both the input image and to the equalized input image by a general histogram equalization method. Since the details and edge information are kept in high pass sub bands, the method concentrates on the approximation sub bands. After obtaining both approximation sub bands, a Singular Value Decomposition (SVD) is applied to the approximation sub bands of the input and the equalized images. The singular value calculated from the input image is weighted by the singular value of the equalized image to obtain an enhanced singular value. Finally, inverse SVD followed by inverse DWT are applied to obtain the enhanced image. A more recent transform domain method uses the Bilateral Filtering (BF) for image enhancement [1]. The input image is decomposed into its approximation and detail layers by a multiscale BF. Finally, the obtained detail layers are added to the original image with a weighted manner to obtain edge enhanced image. Another method is the Remote Sensing Image Enhancement based on the hazy image model [2]. In this method, the commonly used hazy image model [26] is adapted for image enhancement applications. Here, the two unknown parameters of the hazy image model, namely airlight and transmission, are estimated with simple statistical properties of the input image to obtain the enhanced image. A more recent work is based on Robust Guided Filtering [24]. In this method, a robust guided filter described in [27] is applied to the input image and the difference between the original image and filtered image is considered as a detail sub-band as in DWT. The detail sub-bands are amplified and added to the original image to obtain the final enhanced image. Although they show a better performance, the methods in this group suffer from blocking artifacts or, in some cases, they are unable to enhance the image globally [22]. The overall performance of transform domain methods is better than the histogram modification methods. Moreover, the performance of this group of methods is significantly better for high resolution images and images containing both low and high scale details. There are hybrid methods combining histogram and transform methods. One hybrid method is based on a Regularized Histogram Equalization and Discrete Cosine Transform (RHE-DCT) [25]. In this technique, first a global enhancement is applied to the input image by a Regularized Histogram Equalization (RHE). Here, the equalization is made by using the sigmoid function. After obtaining the equalized image, Discrete Cosine Transform (DCT) is applied to the equalized image to obtain DCT coefficients. After this, the coefficients are modified to locally improve the contrast of the image. Finally, inverse DCT is applied to obtain the enhanced image.

In addition to all these methods, a hybrid algorithm combining [1] and HIM [2] methods has been proposed. In this proposed hybrid algorithm, the BF method described above is applied to the image to obtain a global enhanced image. Then, the HIM method is applied block by block to this globally enhanced image to obtain a local enhancement.

2. Remote sensing image enhancement methods

The quality of remote sensing images depends upon numerous factors such as noise, illumination or equipment conditions during the acquisition procedure [28]. The data obtained by optic sensors (multispectral, hyperspectral, panchromatic sensors) are degraded by atmospheric effects and instrumental noises, namely thermal (Johnson) noise, quantization noise and shot (photon) noise which cause corruption in the spectral bands by varying degrees [29]. On the other hand, SAR images (radar sensors), which offer many benefits such as working 7/24 and in all weather conditions, suffer from multiplicative speckle noise [28].

These degradations reduce the contrast in the resulting images and can highly affect human perception or the accuracy of computer assisted applications [25].

Thus, contrast enhancement, besides noise removal, constitutes a primary step for various applications of remote sensing image processing for better information representation and visual perception.

2.1 Adaptive gamma correction with weighting distribution (AGCWD)

In this method, a weighting distribution of the original histogram of the input image is obtained followed by Gamma correction.

First an Adaptive Gamma correction is made to the input image as:

$$T(l) = l_{max} \left(\frac{l}{l_{max}} \right)^{\gamma} = l_{max} \left(\frac{l}{l_{max}} \right)^{1-F(l)} \tag{1}$$

Here, l is the intensity value of the current pixel and l_{max} is the maximum intensity value of the input image. γ is a varying adaptive parameter which is equal to $1 - F(l)$, and $F(l)$ is the cumulative distribution function. The reason to use cumulative distribution function for adaptive Gamma correction is to guarantee the Gamma parameter to follow the changes between the pixels of the image.

In order to avoid the adverse effects, a weighting distribution function is used so as to slightly modify the histogram as follows:

$$f(l) = f_{max} \left(\frac{f(l) - f_{min}}{f_{max} - f_{min}} \right)^{\alpha} \tag{2}$$

Here, α is the adjustment parameter, f is the probability density function and f_{max} and f_{min} are the maximum and minimum of the f. Using (2), the modified cumulative distribution function F is evaluated by:

$$F_\omega(k) = \frac{\sum_{l=0}^{k} f_\omega(l)}{\sum f_\omega} \tag{3}$$

where

$$\sum f_\omega = \sum_{l=0}^{l_{max}} f_\omega(l) \tag{4}$$

Finally, the Gamma parameter of (1) is modified as:

$$\gamma = 1 - F_\omega(l) \tag{5}$$

The modified Gamma parameter and Eq. (1) is used to obtain the enhanced image.

2.2 Discrete wavelet transform and singular value decomposition based method (DWT-SVD)

In this method, a combination of discrete wavelet transform (DWT) and singular value decomposition (SVD) are used for enhancement purposes. In the classical one-dimensional (1D) DWT, the input signal is decomposed into its low (L) and high (H) frequency components. In order to perform a two-dimensional (2D) transform, the 1D DWT is applied to the row of the images followed by the columns of the image, or vice versa. After applying the 2D DWT, four different subbands are obtained, namely LL, LH, HL, and HH. The approximation subband LL contains low frequency components, while the diagonal subband HH contains high frequency components for both rows and columns of the image. The horizontal and vertical subbands LH and HL contains low frequency component for the rows and high frequency components for the columns and vice versa, respectively.

SVD is used to decompose a matrix into two orthogonal square matrices (U and V) and a diagonal matrix containing the singular values (Σ) as shown:

$$I = U_I \Sigma_I V_I^T \tag{6}$$

The enhancement method firstly applies a general histogram equalization to the input image I to obtain equalized image $\tilde{I}$. Then discrete wavelet transform is applied to both the input and equalized images so as to obtain the subbands LL_I, LH_I, HL_I, HH_I and $LL_{\tilde{I}}$, $LH_{\tilde{I}}$, $HL_{\tilde{I}}$, $HH_{\tilde{I}}$, respectively.

Since the rough information about the images are present in the LL subbands, SVD is applied to these subbands to obtain the singular values. As aforementioned, the singular values contain the intensity of the image. Therefore, equalization is made for the singular values. Here, the Σ components of the LL_I and $LL_{\tilde{I}}$ is weighted as to obtain a correction coefficient (ξ).

$$\xi = \frac{max\left(\Sigma_{LL_{\tilde{I}}}\right)}{max\left(\Sigma_{LL_I}\right)} \tag{7}$$

where $\Sigma_{LL_{\tilde{I}}}$ is the singular value matrix of the equalized image derived from its $LL_{\tilde{I}}$ subband and Σ_{LL_I} is the singular value matrix of the input image obtained from its LL_I subband. After determining the correction coefficient, the corrected singular value matrix Σ is obtained as:

$$\Sigma = \xi \Sigma_{LL_I} \tag{8}$$

Here, Σ is the corrected singular value matrix. The new LL subband is constructed as:

$$LL = U_{LL_I} \Sigma V_{LL_I}^T \tag{9}$$

After constructing the new LL subband, the enhanced image is obtain by performing the inverse DWT to this new LL subband and detail subbands of the original image.

2.3 Regularized histogram equalization and discrete cosine transform based method (RHE-DCT)

This method basically consists of two steps: Regularized Histogram Equalization (RHE) followed by Discrete Cosine Transform (DCT). The first one performs a global contrast enhancement and the second one enhances the local contrast.

RHE aims to perform a histogram equalization to the input image by a regularized manner as:

$$f(k) = s(k)(1 + h(k)) \tag{10}$$

Here $f(k)$ is the probability density function of the equalized histogram, $h(k)$ is the normalized histogram of the input image, $s(k)$ is the sigmoid function defined as:

$$s(k) = \frac{1}{1 + e^{-(k-1)}} - \frac{1}{2} \tag{11}$$

By this modification, the minimum value of the equalized image is assured to be equal to 0. The $f(k)$ obtained is normalized as:

$$f(k) \leftarrow \frac{f(k)}{\sum_{t=1}^{K} f(t)} \tag{12}$$

Here K is the number of the gray levels. The cumulative distribution function $F(k)$ is obtained as:

$$F(k) = \sum_{t=1}^{K} f(t) \tag{13}$$

and new gray levels are evaluated as:

$$y(k) = \left[F(k) \left(y_{max} - y_{min} \right) + y_{min} \right] \tag{14}$$

Finally the equalized image is obtained by using a standard lookup table based HE procedure to obtain Y_{eq}.

In order to perform a local enhancement, the DCT coefficients of the globally equalized image is used. For this purpose, first the DCT is applied to the equalized image as:

$$C(u,v) = c_h c_\omega \sum_{k=0}^{M-1} \sum_{l=0}^{N-1} Y_{eq}(k,l) \cos \left[\frac{(2k+1)h\pi}{2M} \right] \cos \left[\frac{(2l+1)\omega\pi}{2N} \right] \tag{15}$$

c_h and c_ω are computed by:

$$c_h = \begin{cases} \sqrt{\dfrac{1}{M}}, & h = 0 \\[2mm] \sqrt{\dfrac{2}{M}}, & 1 \leq h \leq M - 1 \end{cases} \tag{16}$$

$$c_\omega = \begin{cases} \sqrt{\dfrac{1}{N}}, & \omega = 0 \\[2mm] \sqrt{\dfrac{2}{N}}, & 1 \leq \omega \leq N - 1 \end{cases} \tag{17}$$

The lower absolute values of C should be adjusted to perform local enhancement while higher values should be maintained to avoid drastic changes. By this way new DCT coefficients are obtained as:

$$D'(h,w) = \begin{cases} D(h,w), & D(h,w) > 0.01D(0,0) \\ \alpha D(h,w), & D(h,w) \leq 0.01D(0,0) \end{cases} \tag{18}$$

Here α is the adjustment parameter and is automatically determined as:

$$\alpha = 1 + \sqrt{\left(std\left(Y_{global} \right) \right) - std(X) / \left(2^B - 1 \right)} \tag{19}$$

After obtaining the new DCT coefficients, inverse DCT is applied to obtain the final enhanced image.

2.4 Bilateral filtering based method (BF)

This method is basically based on multiscale bilateral filtering. In classical bilateral filtering, the filter output can be determined as:

$$\text{BF}[I] = \left(\frac{1}{W}\right)\sum_{q \in s}G_{\sigma_s}(\|p - q\|)G_{\sigma_r}(\|I_p - I_q\|)I_q \tag{20}$$

where

$$W = \sum_{q \in s}G_{\sigma_s}(\|p - q\|)G_{\sigma_r}(\|I_p - I_q\|) \tag{21}$$

Here σ_s and σ_r are the Gaussian kernels controlling the spatial and range of the input image. I_p is the intensity value of the pixel at location p, I_q is the intensity value of the neighboring pixels within the window S at location q. The difference between the input image and the filter output gives the detail layer of the image.

$$D^1 = I - \text{BF}[I] \tag{22}$$

Here, D^1 is the first detail layer of the image. In order to carry on the decomposition, bilateral filtering is applied again to the filter output. Here, to guarantee the shift invariance, σ_s is doubled and σ_r is halved. In order to obtain level detail layer, two adjacent filter outputs are subtracted as:

$$D^j = \text{BF}^j[I] - \text{BF}^{j-1}[I] \tag{23}$$

Here, j corresponds to the decomposition level.

In order to reconstruct the input image from an L levels of decomposition, one can simply add all detail layers to the final filtering output as:

$$I = \sum_{j=1}^{L}D^j[I] + \text{BF}^L[I] \tag{24}$$

Bilateral filtering based method firstly decomposes the input image by (24).

After obtaining the detail layers for L levels. The details are amplified and added directly to the original image as:

$$I_E = \text{BF}^L[I] + \sum_{j=1}^{L}\omega_j D^j[I] \tag{25}$$

Here, I_E is the enhanced image and ω_j are the weighting factors for the corresponding detail subbands $D^j[I]$.

The parameter determination is very important in order to achieve a good enhancement result. Therefore, σ_r, σ_s, and S parameters of the bilateral filter, as well as the decomposition level and weights have to be determined. To achieve this, a comparison between the enhancement results obtained by differing parameter are made. As a result of this comparison, σ_r is chosen as 0.6, σ_s is chosen as 1.8, S is chosen as a window sized 5×5, the decomposition level is chosen as 4, and the weights $(\omega_1, \omega_2, \omega_3, \omega_4)$ are chosen as 2 [1].

2.5 Adaptive cuckoo search based enhancement algorithm (ACSEA)

In this method, the image enhancement is performed by optimizing a predefined enhancement kernel [30]. The enhancement process of ACSEA is given below:

$$I_{E(i,j)} = \left(\mu^L_{(i,j)}\right)^\alpha + F^e_{(i,j)}\left(I_{i,j} - c^L_{\mu_{(i,j)}}\right) \tag{26}$$

where

$$F^e_{(i,j)} = k\frac{\mu^G}{\sigma^L_{(i,j)} + b} \tag{27}$$

Here, $F_{((i,j))^e}$ is calculated by the mean value and standard deviation of the image and called as the image enhancement function. (i,j) is the location of the current pixel. $\sigma_{((i,j))^L}$ is the local standard deviation and $\mu_{((i,j))^L}$ is the local mean value calculated in a window sized $N \times N$ centered at (i,j), while μ^G is the global mean value. The method focuses on optimizing the parameters (a,b,c,k), where $0 \le a \le 1.5$, $0 \le b \le 0.5$, $0 \le c \le 1$, and $0.5 \le k \le 1.5$.

In order to optimize the enhancement formula given in (26), a chaotic initialization is made and an objective fitness function is used as given below:

$$F(I_E) = \log\left(\log\left(E(I_E^8)\right) + e\right)\frac{N_e\left(I_E^8\right)}{MN}e^{H(0)} \tag{28}$$

where

$$I_E^8 = \sqrt{\left(\nabla_x I_E\right)^2 + \left(\nabla_y I_E\right)^2} \tag{29}$$

In (28), $E(.)$ is the expected value operator and $H(.)$ is the entropy operator. In (29), ∇_x and ∇_y are the gradients. I_E^8 is the Sobel edge detected image.

In order to optimize the enhanced image (I_E), the objective function given in (29) is optimized with a chaotic initialization so as to obtain the best enhancement result.

2.6 Hazy image model based enhancement (HIM)

This method is based on the commonly used hazy image model [26, 31].

$$I = Jt + A(1 - t) \tag{30}$$

where I is the input image, A is the airlight coefficient, t is the transmission map and J is the haze free image. In order to obtain haze free image J, A and t have to be estimated.

For dehazing purposes airlight coefficient is generally estimated from the brighter pixels of the input image. For enhancement, instead of the brighter pixels the mean of the image is assumed to be the airlight coefficient [2].

$$A = 1/KL \sum_{k=1}^{K}\sum_{l=1}^{L}I(k,l) \tag{31}$$

I is the input image with dimensions of $K \times L$ and $I(k,l)$ is the intensity value of the pixel at location (k,l). In general, in dehazing algorithms, the transmission map is estimated using the airlight coefficient and normalized input image. The normalized image is obtained by estimated airlight coefficient. Following the similar manner, this method also normalizes the image with the estimated airlight and estimate the transformation as:

$$t = 1 - \omega \left(\frac{1}{A} \right) \tag{32}$$

Here, ω is an arbitrary coefficient. The coefficient can be determined as the standard deviation (σ) of the input image [2]. Finally, the enhanced image is obtained by simply taking out J out of (30) as:

$$J = \frac{I - A(1 - t)}{t} \tag{33}$$

2.7 Robust guided filtering based method (SDF)

This method uses the Robust Guided Filtering described in [27] which uses two guidance images namely dynamic guidance and static guidance. In order to perform Robust Guided Filtering, the following cost function should be minimized:

$$\in_u = \sum_i c_i \left(u_i - f_i \right)^2 + \lambda \Omega(u,g) \tag{34}$$

Here, f is the input image, u is the dynamic guidance and g is the static guidance. λ is the regularization parameter and $c_i \leq 0$ is the confidence level. The regularizer $\Omega(u,g)$ can be defined as [27].

$$\Omega(u,g) = \sum_{i,j \in N} \phi_\mu \left(g_i - g_j \right) \varphi_v \left(u_i - u_j \right) \tag{35}$$

where

$$\varphi_v(x) = \frac{1 - \varphi_v(x)}{v} \quad \text{and} \quad \phi_\mu(x) = e^{-\mu x^2} \tag{36}$$

N is the neighborhood size which is 8×8, while μ and v are parameters controlling the smoothness level.

In order to perform image enhancement, a multi-scale decomposition based on Robust Guided Filtering similar to the multi-scale bilateral filtering is proposed in [1]. The filtering output is considered as the first approximation layer of the original image as:

$$A_1 = \text{SDF}[I] \tag{37}$$

Here, I is the input image, A_1 is the first level approximation layer and SDF operator stands for Robust Guided Filtering. In order to obtain further levels of approximation layers, SDF is applied to previous approximation layer as:

$$A_l = \text{SDF}[A_{l-1}] \tag{38}$$

with initial value $A_1 = I$. The difference between two adjacent approximation layers give the detail layer of the corresponding level as:

$$D_l = A_l - A_{l-1} \qquad (39)$$

One can obtain the original image by simply adding the detail layers to the final level approximation layer.

$$I = \sum_{j=1}^{L} D_j + A_L \qquad (40)$$

SDF based enhancement firstly decomposes the input image by using (40).
After obtaining the detail layers, the details are amplified and added directly to the original image as:

$$I = \sum_{j=1}^{L} \omega_j D_j + A_L \qquad (41)$$

The decomposition level and weights are determined by comparing different number of levels and weights. The best results for different images are applied for all images. Therefore, the decomposition level is chosen as 4. Moreover, the weights $(\omega_1, \omega_2, \omega_3, \omega_4)$ are chosen as 2 [24].

2.8 Hybrid bilateral filtering and hazy image model method (BF-HIM)

The BF based enhancement method [1] has a good enhancement, however the color distortion is present, whereas the HIM method [2] has a good color preservation with a lower enhancement performance. Therefore, a hybrid method combining these two methods can be a good candidate to obtain a good performance for enhancement along with a good color preservation.

The hybrid method first applies the multi-scale bilateral filtering given in (24) to the input image to obtain the bilateral filtering outputs and detail layers. Since, we will add the HIM model, the decomposition level is chosen as 2. Then, the detail layers are amplified as given in (25) to obtain the prior enhancement result. The prior enhanced image is divided into non overlapping blocks. HIM method given above is applied to these blocks separately to perform a local enhancement. Finally, the enhanced blocks are combined to construct the final enhancement result.

Here, the choice of the block size is important. The lower block size is expected to have a better local enhancement result. Therefore, the block size is chosen as 3×3.

3. Evaluation criteria

It is possible to determine the performance of an image enhancement method visually. However, a visual conclusion may not be objective. Therefore, in order to make objective comparisons, evaluation criteria has been developed. Here, the choice of criteria is also important. It is already known that every criterion can give an idea about one property of the resulting image. Therefore, criteria for different properties of the image should be used. Moreover, since each criterion gives an idea for a certain property of the image, all criteria should be considered together to have an overall idea of the image. The criteria presented below gives an idea for the performance of the enhancement methods, however all should be considered together and along with visual results.

3.1 Contrast gain (CG)

The first criteria to measure the performance of enhancement method is Contrast Gain (CG) [32]. This criterion focuses on the contrast improvement of the image as follows:

$$CG = \frac{C(Y)}{C(X)} \tag{42}$$

where C is average of the local Michelson contrast, which is calculated for 3x3 sized windows within an image and given as:

$$C = \frac{max - min}{max + min} \tag{43}$$

The higher CG value indicates that the contrast improvement is better.

3.2 Enhancement measurement (EME)

This criterion also considers the contrast improvement within the enhanced image and defined by following [22]:

$$EME_{a,k_1,k_2}(\varphi) = \frac{1}{k_1 k_2} \sum_{l=1}^{k_1} \sum_{k=1}^{k_2} \alpha \left(\frac{I_{max}^{k,l}(\varphi)}{I_{min}^{k,l}(\varphi)+c} \right)^\alpha \ln \frac{I_{max}^{k,l}(\varphi)}{I_{min}^{k,l}(\varphi)+c} \tag{44}$$

Here the image I is split into $k_1 \times k_2$ sized blocks. $I_{max}^{(k,l)}$ and $I_{min}^{(k,l)}$ are the maximum and minimum values within the block, while c is a small constant to avoid division by zero. $EME_{a,k_1,k_2}(\varphi)$ is called the Enhancement Measurement of Entropy with respect to transform φ.
The higher EME value indicates that the contrast improvement is better.

3.3 Discrete entropy (DE)

Discrete entropy of an image can be evaluated as:

$$DE = -\sum_{k=1}^{K} p(x_k) \log p(x_k) \tag{45}$$

Here, $p(x_k)$ is the probability of the pixel (x_k). The higher value of DE indicates that a smoother distributed histogram is obtained, which may indicate that the contrast is higher.

3.4 Absolute mean brightness error (AMBE)

Absolute Mean Brightness Error (AMBE) [18] is an error function calculated between image X and image Y as:

$$\text{AMBE} = \frac{1}{MN} \sum_{m=1}^{M} \sum_{n=1}^{N} |X(m,n) - Y(m,n)| \tag{46}$$

Here, M and N are the dimensions of the images and (m, n) is the pixel location. The lower AMBE value indicates that, the brightness preservation is better.

4. Experimental setup

The enhancement described above have been applied to several images. Comparisons of the methods are made both visually and quantitatively. Before applying the enhancement methods, the parameters for each method is determined.

4.1 Visual comparison

Visual comparisons are performed for different images and they are available online.[1]

The first image used for comparison is a tank image taken by a digital imaging system as shown in **Figure 1(a)**. **Figure 1(b)–(i)** show the enhancement results obtained by AGCWD, DWT-SVD, RHE-DCT, ACSEA, BF, HIM, SDF, and BF-HIM methods, respectively. In order to demonstrate closely, the zoomed version of the area inside the red square is given inside the green square. As seen in **Figure 1(b)**, AGCWD method has improved the contrast, however the color preservation of the method is not good. The contrast improvement of DWT-SVD seems to be low, as seen in **Figure 1(c)**. Even though the color preservation seems to be good, the edge information is lost as seen in the zoomed area. RHE-DCT method, shown in **Figure 1(d)**, has a good contrast improvement, however the color preservation is not good. The edge enhancement of RHE-DCT is better than AGCWD and DWT-SVD methods. ACSEA method in **Figure 1(e)** demonstrates a better color preservation, however the contrast improvement is not as good as the other methods. Moreover, the edge enhancement is lower than RHE-DCT method. As seen in **Figure 1(f)**, BF method preserves the color like as the ACSEA method and has a good edge enhancement performance. **Figure 1(g)** shows that HIM method has a good color preservation capability. However, the edge enhancement performance is not good compared to the BF method. SDF method, given in **Figure 1(h)** has a very good edge enhancement performance, but the color preservation is lower than ACSEA, HIM and BF methods. As demonstrated in **Figure 1(i)**, the hybrid BF-HIM method preserve the colors closer to the ACSEA and BF methods and enhances the edge information better than the former methods.

The second image used for comparison is an aerial image taken by a digital imaging system mounted on an air vehicle as shown in **Figure 2(a)**. **Figure 2(b)–(i)** show the enhancement results obtained by AGCWD, DWT-SVD, RHE-DCT, ACSEA, BF, HIM, SDF, and BF-HIM methods, respectively. In order to make a closer look, a zoomed version of the area inside the red square is given in the green square.

As seen in **Figure 2(b)**, AGCWD method has improved the contrast, however the color preservation of the method is not good. The car within the zoom area is visible. The contrast improvement of DWT-SVD is better than AGCWD method, as seen in **Figure 2(c)**. The color preservation is lower than AGCWD method and the

---

[1] http://sipi.usc.edu/database/

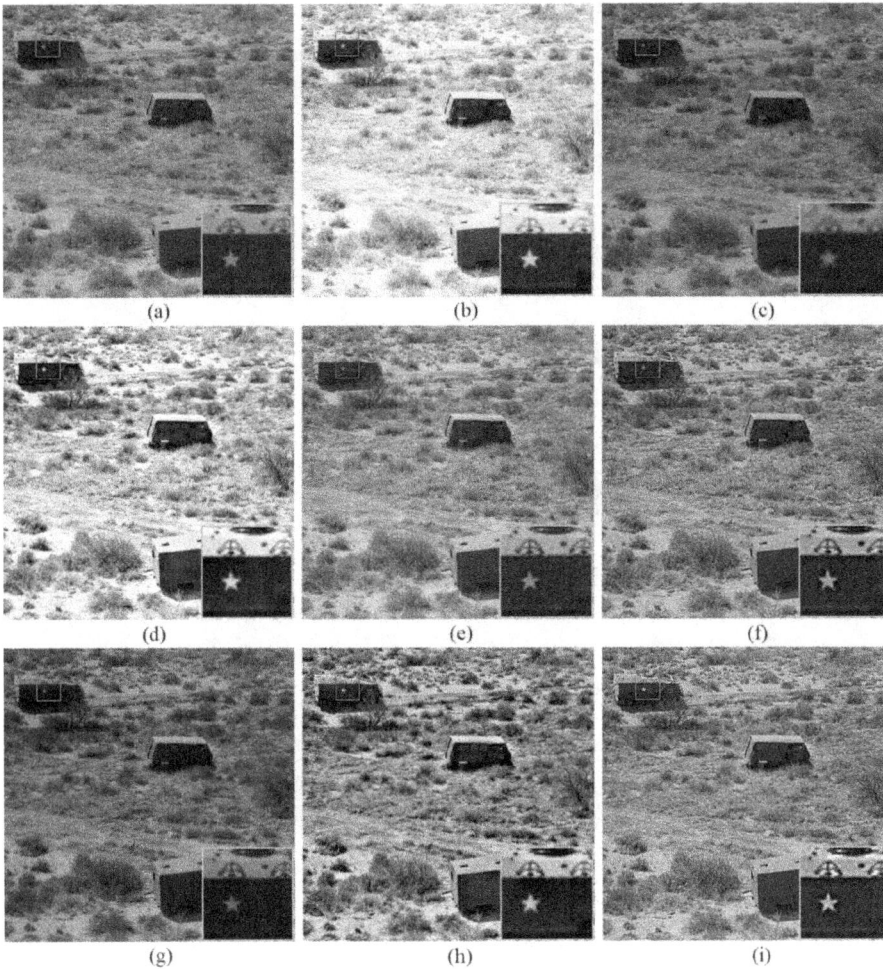

Figure 1.
(a) Input image, enhancement results for (b) AGCWD (c) DWT-SVD, (d) RHE-DCT, (e) ACSEA, (f) BF, (g) HIM, (h) SDF, and (i) BF-HIM methods.

visibility of the car in the zoomed area is not as good as AGCWD method. RHE-DCT method, shown in **Figure 2(d)**, has a good contrast improvement and better color preservation than AGCWD and DWT-SVD methods. The edge enhancement of RHE-DCT is closer to the AGCWD method as seen in the zoomed area. ACSEA method in **Figure 2(e)** demonstrates a better color preservation, however the contrast improvement is not as good as the other methods. Moreover, the edge enhancement is lower than RHE-DCT method. As seen in **Figure 2(f)**, BF method preserves the color like as the ACSEA method and has a better edge enhancement performance than RHE-DCT methods, as seen in the zoomed area. **Figure 2(g)** shows that HIM method has a good color preservation capability. However, the edge enhancement performance is not good compared to the BF method. SDF method, given in **Figure 2(h)** has a very good edge enhancement performance, but the color preservation is lower than ACSEA, HIM and BF methods. As demonstrated in **Figure 2(i)**, the hybrid BF-HIM method preserve the colors closer to the HIM method. A closer look demonstrates that the edge improvement is better than the former methods.

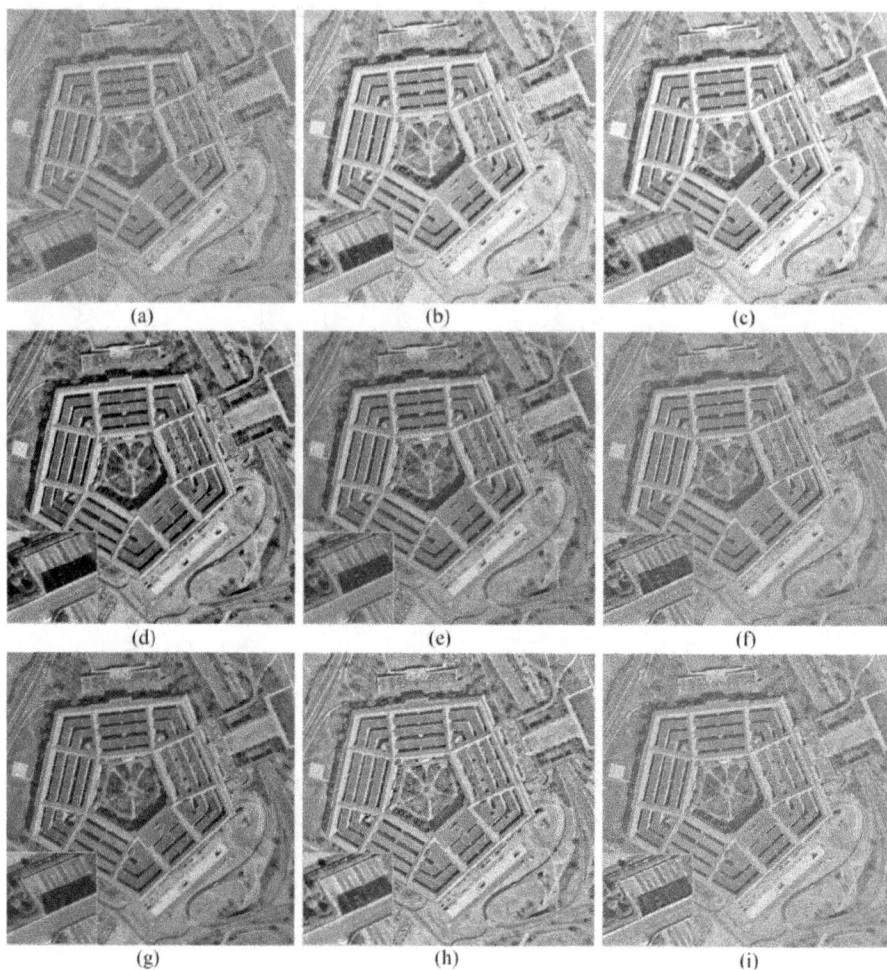

Figure 2.
(a) Input image, enhancement results for (b) AGCWD (c) DWT-SVD, (d) RHE-DCT, (e) ACSEA, (f) BF, (g) HIM, (h) SDF, and (i) BF-HIM methods.

The final image used for comparison is an aerial image of an area containing harbor and airport taken by a digital imaging system mounted on an air vehicle as shown in **Figure 3(a)**. **Figure 3(b)**–**(i)** show the enhancement results obtained by AGCWD, DWT-SVD, RHE-DCT, ACSEA, BF, HIM, SDF, and BF-HIM methods, respectively. For a closer look, the area shown in red square is zoomed and given within the green square.

As seen in **Figure 3(b)**, AGCWD method has improved the contrast, however the color preservation of the method is not good. Moreover, the edge information is lost as seen in the zoomed area. The contrast improvement of DWT-SVD seems to be low, as seen in **Figure 3(c)**. Even though the color preservation seems to be good, the edges have not been improved as seen in the zoomed area. RHE-DCT method, shown in **Figure 3(d)**, has a good contrast improvement, and the color preservation is good. The edge enhancement of RHE-DCT is better than AGCWD and DWT-SVD methods. ACSEA method in **Figure 3(e)** demonstrates a good color preservation, and a fine contrast improvement. Moreover, the edge improvement seems to be better than RHE-DCT method. As seen in **Figure 3(f)**, BF method

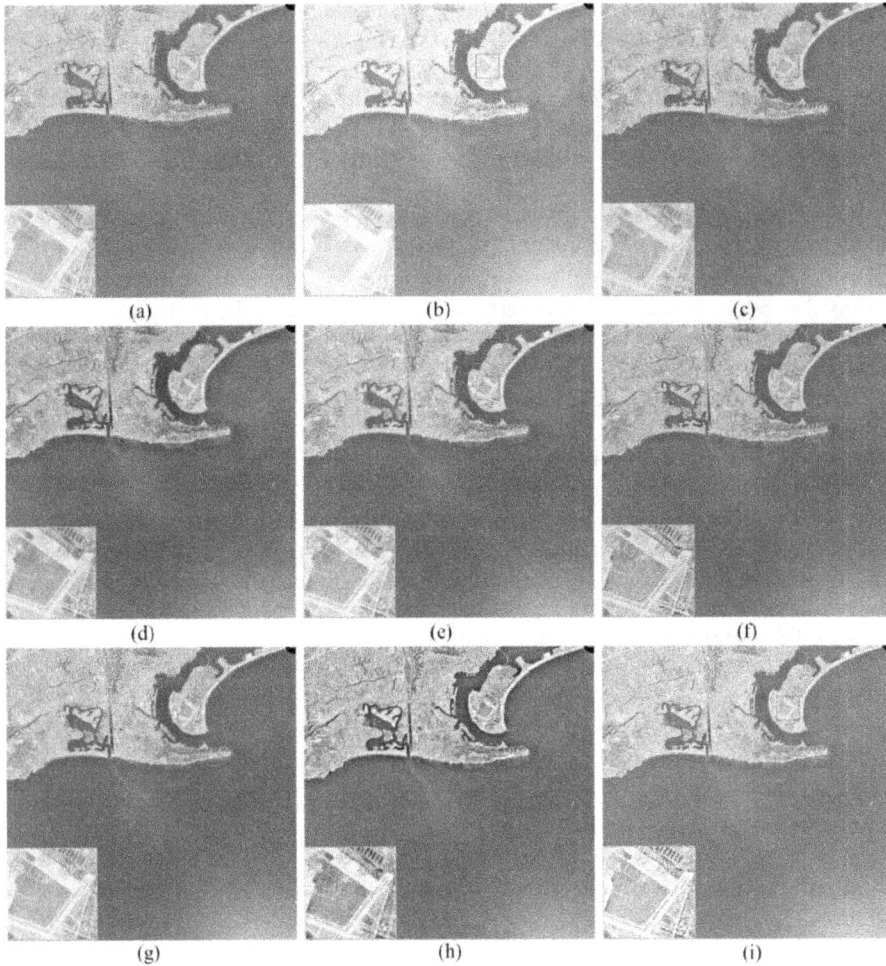

Figure 3.
(a) Input image, enhancement results for (b) AGCWD (c) DWT-SVD, (d) RHE-DCT, (e) ACSEA, (f) BF, (g) HIM, (h) SDF, and (i) BF-HIM methods.

preserves the color better than ACSEA method and has a good edge enhancement performance. **Figure 3(g)** shows that HIM method has a good color preservation capability. However, the edge enhancement performance is not good enough compared to the BF method. SDF method, given in **Figure 3(h)** has a very good edge enhancement performance, but the color preservation is lower than ACSEA, HIM and BF methods. As demonstrated in **Figure 3(i)**, the hybrid BF-HIM method preserve the colors closer to the ACSEA methods and enhances the edge information better than the former methods.

For an objective visual evaluation, the profiles of the horizontal lines given in **Figure 1(a)**, **Figure 2(a)** and **Figure 3(a)** are constructed for the enhancement methods, and the drawn profiles for the original image are given along with enhancement methods are given in **Figure 4(a)–(c)**, respectively.

According to **Figure 4(a)**, DWT-SVD and ACSEA methods cannot follow the changes which means the details of the image are lost for these methods. BF-HIM method can follow the changes better. Moreover, BF-HIM method have increased the intensity range more compared to the other methods, which indicates that the contrast improvement is better for BF-HIM method.

According to **Figure 4(b)**, all three methods seem to follow the pattern of the original image properly, in general. ACSEA method has lost the pattern in some parts. SDF and BF-HIM methods have followed the pattern better than ACSEA method. Moreover, BF-HIM seems to have a slightly wider range than SDF method as well.

According to **Figure 4(c)**, all three methods seem to follow the pattern of the original image properly. AGCWD method have increased the intensity values in general, which results in a brighter region. By this way, the contrast improvement is not good enough. Similarly, HIM method have decreased the intensity values, which results in a darker region. BF-HIM method has improved the contrast better than the other methods.

Therefore, according to the visual comparisons, the higher the detail level is within the image, the better the results for methods like AGCWD and RHE-DCT are, as expected, since both methods use histogram modification.

It can also be concluded that histogram modification methods like AGCWD and RHE-DCT methods have a good performance, if the resolution is low (**Figure 1**) or/ and the input image contains higher-scale edge information (**Figure 2**), while transform domain methods are generally better for high resolution images or/and images containing small-scale details. Also, transform domain methods seem to have a solid performance for high-scale details.

In addition to this, considering all aspects of the resulting images, in terms of color preservation, contrast improvement, and edge enhancement, the BF, and hybrid BF-HIM methods seem to have better results. Moreover, hybrid BF-HIM method seems to be the best method when looking at all three aspects.

4.2 Quantitative comparison

In order to perform an objective comparison, the criteria aforementioned are evaluated for enhancement results obtained by the methods, the visual results of which are given in **Figures 1–3**. The quantitative results are provided in **Tables 1–4** where the best results are emphasized in bold. The first criterion used for comparison is the Contrast Gain (CG). **Table 1** shows the CG values obtained for AGCWD, DWT-SVD, RHE-DCT, ACSEA, BF, HIM, SDF, and BF-HIM methods.

According to **Table 1**, for **Figure 1**, the best score is obtained by SDF followed by BF-HIM method. For **Figure 2**, the best score is obtained by RHE-DCT method followed by BF-HIM method. For **Figure 3** is achieved by hybrid BF-HIM method, followed by SDF method. Therefore, it is possible to say that RHE-DCT method has a better contrast gain for images containing high-scale details like **Figure 3**.

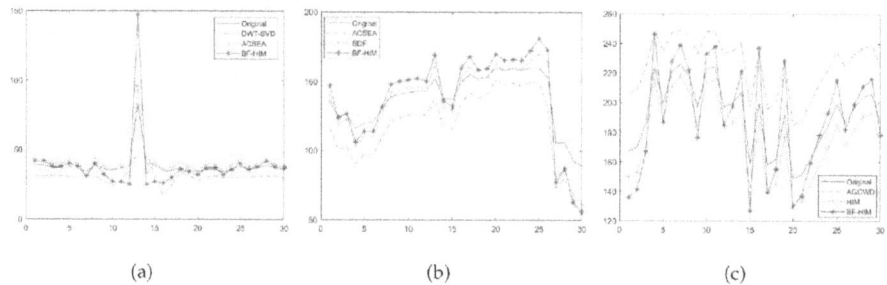

(a)	(b)	(c)

Figure 4.
Drawn profiles for input and enhanced images for (a) Figure 1, (b) Figure 2, (c) Figure 3.

The second criterion used for comparison is the Enhancement Measurement (EME). **Table 2** shows the EME values obtained for AGCWD, DWT-SVD, RHE-DCT, ACSEA, BF, HIM, SDF, and BF-HIM methods.

According to **Table 2**, for **Figure 1** the best EME scores are obtained by BF-HIM method, followed by SDF method. For **Figure 2**, the best score is obtained by DWT-SVD method followed by ACSEA method. For **Figure 3**, the best EME scores are obtained by BF-HIM method for followed by SDF method. Therefore, it is possible to say that DWT-SVD method has a better enhancement performance for images containing high-scale details, likeas in **Figure 2**.

The third criterion used for comparison is the Discrete Entropy (DE). **Table 3** shows the DE values obtained for AGCWD, DWT-SVD, RHE-DCT, ACSEA, BF, HIM, SDF, and BF-HIM methods.

According to **Table 3**, for **Figure 1**, the best score is obtained by RHE-DCT method followed by BF-HIM method. For **Figure 2**, the best score id obtained by RHE-DCT method followed by SDF method. As it is seen in DE values, the higher the scale of detail is within the images, the higher is the performance of RHE-DCT method. BF-HIM method has better DE values for **Figure 3**, and has a close score to RHE-DCT method for **Figures 1** and **2**.

The fourth criterion used for comparison is the Absolute Mean Brightness Error (AMBE). **Table 4** shows the AMBE values obtained for AGCWD, DWT-SVD, RHE-DCT, ACSEA, BF, HIM, SDF, and BF-HIM methods.

According to **Table 4**, for **Figure 1**, the best score is obtained by ACSEA method followed by BF-HIM method. For **Figure 2**, the best score is obtained by BF-HIM method followed by BF method. For **Figure 3**, the best value is obtained by BF-HIM

Method	AGCWD	DWT-SVD	RHE-DCT	ACSEA	BF	HIM	SDF	BF-HIM
Figure 1	1.0600	0.7468	1.6442	1.4369	1.9099	1.3171	**2.2560**	2.0457
Figure 2	1.5302	1.1635	**2.9214**	1.6665	1.8910	1.6529	2.0120	2.0517
Figure 3	1.0555	0.5505	1.3110	1.7592	1.9522	1.1654	2.0217	**2.0789**

Table 1.
CG values obtained for the enhancement methods.

Method	AGCWD	DWT-SVD	RHE-DCT	ACSEA	BF	HIM	SDF	BF-HIM
Figure 1	1.47	1.91	327.05	4.11	329.11	6.00	384.70	**407.07**
Figure 2	1.29	**7.92**	5.86	7.44	1.81	1.66	4.94	6.95
Figure 3	2.59	5.01	4.41	1.35	6.91	2.00	7.98	**9.29**

Table 2.
EME values (10^4) obtained for the enhancement methods.

Method	AGCWD	DWT-SVD	RHE-DCT	ACSEA	BF	HIM	SDF	BF-HIM
Figure 1	6.5601	6.8245	**7.8527**	7.3600	7.3778	6.4493	7.6648	7.6723
Figure 2	6.6672	6.8799	**7.6940**	7.0889	7.1340	6.7327	7.4061	7.2010
Figure 3	5.9251	5.7998	6.3719	6.4604	6.4508	5.9944	6.6397	**6.6682**

Table 3.
DE values obtained for the enhancement methods.

Method	AGCWD	DWT-SVD	RHE-DCT	ACSEA	BF	HIM	SDF	BF-HIM
Figure 1	0.2798	0.0657	0.1871	**0.0149**	0.0652	0.1014	0.0692	0.0459
Figure 2	0.0904	0.1083	0.1145	0.1503	0.0364	0.1265	0.0355	**0.0254**
Figure 3	0.1121	0.0331	0.0811	0.0629	0.0713	0.0651	0.0697	**0.0326**

Table 4.
AMBE values obtained for the enhancement methods.

method followed by DWT-SVD method. Since, AMBE is the error between the original image and enhanced image, the smaller value of AMBE indicates better color preservation. This criterion does not give an idea about enhancement performance.

As a result, even though the visual comparison may give the observer an idea about the enhancement performance, quantitative comparison has to be made to obtain a more objective conclusion. Here, the choice of the quantitative criterion is also important. As it is known, each criteria indicates different aspects for the resulting images. For instance, CG gives an idea about the contrast improvement, while AMBE is about the color preservation. If the aim is to compare the overall performance for the methods aforementioned, all criteria should be considered all together. Thus, the quantitative comparisons, as well as the visual comparisons demonstrate that the hybrid methods combining different methods like BF-HIM result in better enhanced images.

5. Conclusion

The use of image enhancement methods which improve the contrast and edge information of the image is vital for remote sensing applications. In this work, different remote sensing image enhancement methods based on histogram modification techniques (HE, AGCWD) and transform domain methods (DWT-SVD, ACSEA, RHE-DCT, BF, HIM, and SDF) have been reviewed. The resulting images have been compared visually and quantitatively. For quantitative comparison, several image quality criteria have been used. The resolution and the detail scales of the image affects the performance of the enhancement methods. For instance, the detail scales of the input image affect the performance of RHE-DCT and AGCWD methods deeply. Since both methods are histogram modification methods, even though RHE-DCT also uses a transformation, it can be concluded that histogram modification based methods are better if there are higher-scale details within the image or if the image has a lower resolution. The transform domain methods have a better performance for the images with low-scale details, but also the results of these methods are very solid compared to the histogram based methods for high-scale details, as well.

Another contribution of this work is to introduce a hybrid method, which combines the bilateral filtering with hazy image model. The visual and quantitative results demonstrate that using hybrid methods have a superior performance to the methods applied separately. Therefore, future research on remote sensing image enhancement should focus on hybrid methods.

Author details

Nur Huseyin Kaplan[1*], Isin Erer[2] and Deniz Kumlu[3]

1 Electrical and Electronic Engineering Department, Erzurum Technical University, Erzurum, Turkey

2 Electronics and Communication Department, Istanbul Technical University, Istanbul, Turkey

3 Naval Research Center, Turkish Naval Forces, Istanbul, Turkey

*Address all correspondence to: huseyin.kaplan@erzurum.edu.tr

IntechOpen

References

[1] Kaplan, N.H, Erer, I, Gulmus, N. Remote sensing image enhancement via bilateral filtering. In: Proceedings of the 8th International Conference on Recent Advances in Space Technologies (RAST17), 19-22 June 2017. Istanbul, Turkey: IEEE;2017. p.139–142.

[2] Kaplan, N.H. Remote sensing image enhancement using hazy image model. Optik - International Journal for Light and Electron Optics. 2018;155: 139–148.

[3] Arici, T, Dikbas, S, Altunbasak, Y. A histogram modification framework and its application for image contrast enhancement. IEEE Transactions on Image Processing, 2009;18(9):1921–1935.

[4] Huang, S.C, Cheng, F.C, Chiu, Y.S. Efficient contrast enhancement using adaptive gamma correction with weighting distribution. IEEE Transactions on Image Processing. 2013; 22 (3):1032–1041.

[5] Hanmandlu, M, Jha, D. An optimal fuzzy system for color image enhancement. IEEE Transactions on Image Processing. 2006; 15(10):2956–2966.

[6] Dhnawan, A.P., Buelloni, G., Gordon, R. Enhancement of mammographic features by optimal adaptive neighborhood image processing. IEEE Trans. Med. Imaging. 1986;5:8–15.

[7] Beghdadi, A, Negrate,A.L. Contrast enhancement technique based on local detection of edges. Computer Vision, Graphics, and Image Processing. 1989; 46(2):162–174.

[8] Laxmikant Dash, Chatterji, B.N. Adaptive contrast enhancement and de-enhancement. Pattern Recognition. 1991;24:289–302.

[9] Cheng, H.D, Xu, H.J. A novel fuzzy logic approach to contrast enhancement. Pattern Recognition. 2000;33(5):809–819.

[10] Sherrier, R, Johnson,G. Regionally adaptive histogram equalization of the chest. IEEE Transactions on Medical Imaging. 1987; MI-6(January(1)),1–7.

[11] Sattar, F, Floreby, L, Salomonsson, G, Lovstrom, B. Image enhancement based on a nonlinear multiscale method. IEEE Transactions on Image Processing. 1997;6(6):888–895.

[12] Polesel, A, Ramponi,G, Mathews,V. Image enhancement via adaptive unsharp masking. IEEE Transactions on Image Processing. 2000;9(3):505–510.

[13] Salari, E, Zhang, S. Integrated recurrent neural network for image resolution enhancement from multiple image frames. IEE Proceedings - Vision, Image and Signal Processing. 2003;150 (5):299–305.

[14] Lee, E, Kang, W, Kim, S, Paik, J. Color shift model-based image enhancement for digital multifocusing based on a multiple color-filter aperture camera. IEEE Transactions on Consumer Electronics. 2010;56(2):317–323.

[15] Wong, T, Bouman, C. A, Pollak, I. Image Enhancement Using the Hypothesis Selection Filter: Theory and Application to JPEG Decoding. IEEE Transactions on Image Processing. 2013; 22(3):898–913.

[16] Gonzalez, R.C, Woods, R.E. Digital Image Processing, 2nd ed. Addison-Wesley Longman Publishing;2001.

[17] Kim, Y.-T. Contrast enhancement using brightness preserving bi-histogram equalization. IEEE Trans. Consum. Electron. 1997;43(1):1–8.

[18] Chen, S.-D, Raml, A.R. Contrast enhancement using recursive mean-

separate histogram equalization for scalable brightness preservation. IEEE Trans.Consum. Electron.2003;49(4): 1301--1309.

[19] Celik, T, Tjahjadi, T. Contextual and variational contrast enhancement. IEEE Transactions on Image Processing. 2011; 20(12):3431--3441.

[20] Celik, T. Two-dimensional histogram equalization and contrast enhancement. Pattern Recognition. 2012;45(10):3810--3824.

[21] Demirel, H, Anbarjafari, G. Image Resolution Enhancement by Using Discrete and Stationary Wavelet Decomposition. IEEE Transactions on Image Processing. 2011;20(5):1458–1460.

[22] Agaian, S.S, Silver, B, Panetta, K.A. Transform coefficient histogram-based image enhancement algorithms using contrast entropy IEEE Transactions on Image Processing. 2007;16(3):741--758.

[23] Demirel, H, Ozcinar, C, Anbarjafari, G. Satellite image contrast enhancement using discrete wavelet transform and singular value decomposition. IEEE Geoscience and Remote Sensing Letters. 2010;7(2):333--337.

[24] Kaplan, N.H., Erer, I. Remote Sensing Image Enhancement via Robust Guided Filtering. In: Proceedings of the 9th International Conference on Recent Advances in Space Technologies (RAST19), 11-14 June 2019. Istanbul, Turkey:IEEE;2009. p.447-450.

[25] Fu, X, Wang, J, Zeng, D, Huang, Y, Ding, X. Remote sensing image enhancement using regularized- histogram equalization and DCT. IEEE Geoscience and Remote Sensing Letters. 2015;12(11):2301--2305

[26] Narasimhan, S.G, Nayar, S.K. Contrast restoration of weather degraded images. IEEE Trans. Pattern Anal. Mach. Intell. 2003;25(6):713--724.

[27] Ham, B, Cho, M, Ponce, J. Robust guided image filtering using nonconvex potentials. IEEE Transactions on Pattern Analysis and Machine Intelligence. 2018;40(1):192--207.

[28] Soni,V., Bhandari,A.K.,Kumar, A., Singh, G.K. Improved sub-band adaptive thresholding function for denoising of satellite image based on evolutionary algorithms. IET Signal Processing. 2013;7(8):720--730.

[29] Rasti, B., Scheunders, P., Ghamisi, P., Licciardi, G., Chanussot, J. Noise Reduction in Hyperspectral Imagery: Overview and Application. Remote Sens. 2018;10(3):482.

[30] Suresh, S, Lal, S, Reddy, C.S, Kiran, M.S. A Novel Adaptive Cuckoo Search Algorithm for Contrast Enhancement of Satellite Images. IEEE Journal of Selected Topics in Applied Earth Observations and Remote Sensing. 2017; 10(8):3665–3676.

[31] Kaplan, N.H, Dumlu, A, Ayten, K.K. Single image dehazing based on multiscale product prior and application to vision control. Signal, Image and Video Processing. 2017;11(8):1389–1396.

[32] Shin, J, Park, R.H. Histogram-based locality-preserving contrast enhancement. IEEE Signal Process. Lett. 2015;22 (9):1293--1296.

Feature-Oriented Principal Component Selection (FPCS) for Delineation of the Geological Units Using the Integration of SWIR and TIR ASTER Data

Ronak Jain

Abstract

Geological studies have been performed using the Band Ratios (BR), Relative Band Depth (RBD), Mineral Indices (MI), Principal Component Analysis (PCA), Independent Component Analysis (ICA), lithological and mineral classification techniques from Short-Wave Infrared (SWIR) and Thermal Infrared (TIR) data. The chapter aims to delineate various geological units present in the area using the combination of SWIR and TIR ASTER bands through the Feature-Oriented Principal Component Selection (FPCS) technique. Different BRs and RBDs were applied to map the minerals having Al-OH and Mg-OH compounds with the chemical composition of clay (kaolinite, smectite), mica (sericite, muscovite, illite), ultramafic (lizardite, antigorite, chrysotile), talc, and carbonate (dolomite) from SWIR bands. The MI was used to map quartz-rich, mafic/ultramafic, and carbonate rocks using TIR bands. The BRs, RBDs, and MIs mapped the geological units but every single greyscale image showed a variety of features. To compile these features False Color Composite (FCC) was prepared by the combination of RBDs and MIs in the R:G:B channels which demarked various geological units to a larger extent present in the region. To overcome the limitation, the FPCS technique was applied with the integration of all BRs, RBDs, and MIs. The FPCS technique extracts valuable information from different input bands and shifts the information in the first few bands. The generated eigenvalues and eigenvectors represented the retrieved information in the specific band. The loadings of the eigenvector were used for the selection of the different brands to create the FCC for the delineation of geological strata. The best discrimination was made by the selection of FPCS1, FPCS3, and FPCS6 which differentiated all the geological units like ultramafics, dolomites, thin bands of talc, and muscovite and illite (as phyllite and mica-schist), silica-rich rocks (as quartzite), and granite outcrops.

Keywords: Remote Sensing, Optical data, Feature-Oriented Principal Component Selection, Data integration, Geological studies

1. Introduction

Water (oceans, rivers, lakes, etc.) and land (rocky mountains, hills, peneplain, islands, etc.) are the major components of the Earth's surface out of which only 29%

are occupied by the land surfaces. This 29% land coverage included the forest, desert, mountains, islands, etc. so, a very little amount of land is reserved for geological studies.

Traditional mapping methods are time-consuming and require lots of effort for the preparation of lithological maps, mineral maps, structural maps, etc. But sometimes manually collected data may have errors due to inaccessibility and recording of the data which exaggerate in due course. To avoid these errors and corrections introduced therein an advanced technology came into the picture and is known as Remote Sensing. This technique helps in the mapping of the different litho-units and associated structural features with higher accuracy in a short period as compared to the traditional methods.

Remote Sensing is a tool used for the gathering of the target information without any physical/direct contact with the earth's surface [1–6]. It is a widely used science for the identification and mapping of the various objects/materials present on the earth's crust. The electromagnetic wavelength ranges from 0.38 μm to 100 cm i.e. visible to microwave region [3] is utilized for capturing the information from the earth's surface along with different sensors to capture the EM spectrum's energy [4, 5, 7]. This technique is useful for the monitoring, protection, and management of diverse natural resources and land cover [8]. The geological studies include the demarcation of various lithologies, alteration zones, minerals, and structural features.

Multispectral Remote Sensing is utilized in the domain of geosciences for lithological mapping [9–16], mineral mapping [17–24], identification of the alteration zones related to the base metal mineralization [25–42], structural features as a controlling factor for mineralization [26, 28, 42–46] and mapping for demarcating favorable zones of mineralization [21, 47, 48]. Spectral characteristic absorption features of the rocks and minerals are utilized for the identification and mapping of lithologies and minerals like calcite, dolomite, clay, mica, and ultramafics, etc. The spectral absorption features of minerals vary with chemical composition and the resultant spectral curve varies in shape, depth, position, and asymmetry [49].

Wavelength range from 0.38 to 2.5 μm is utilized for the mapping of the various hydroxyl (Al-OH, Mg-OH), iron oxides (Fe-OH), carbonates (CO_3^{-2}), and sulphates (SO_4^{-2}) bearing minerals like clay, mica, ultramafics, hematite, limonite, dolomite, calcite, etc., due to the presence of characteristic absorption features in the VNIR and SWIR region of the EM spectrum [5, 42, 50–53]. In the case of feldspar, silica-rich rocks, and discrimination between ultramafics and dolomites are possible due to spectral features associated with the TIR region in the wavelength range of 3 to 50 μm [20, 23, 54–58]. The dissimilarities in the spectrum in the TIR spectral-domain arise due to variation in chemical composition and molecular structure.

Geological studies are done with the help of Landsat series, ASTER, Sentinel −2, SPOT, Worldview series, GeoEye, etc. optical remote sensing satellites. They are mainly utilized for the perspective of mineral exploration by using the various methods like band ratio (BR), relative band depth (RBD), Principal Component Analysis (PCA), Independent Component Analysis (ICA), Minimum Noise Fraction (MNF), unsupervised classification (K-means, isodata, etc.), supervised classification (Spectral Angle Mapper, Spectral Feature Fitting, Mixture Tuned Matched Filtering, etc.), machine learning (support vector machine, decision tree, artificial neural network, etc.). Various BR and RBD have been used for the delineation of the different rock outcrops like dolomite, calcite, Iron rich-rocks, ultramafics, epidote, clay and mica minerals, etc. [17, 21–24, 42, 58–60] and mineral prospects by the demarcation & mapping of the associated alteration zones [20, 22, 27, 28, 37, 45, 61]. Lithological mapping of the exposed outcrops and their associated features are demarcated with the help of PCA, ICA, and MNF analyses to govern the mineral potentiality of the outcrops [9, 17, 40, 42, 44, 62, 63]. Different supervised, unsupervised, machine learning and prospectivity mapping algorithms were applied to the optical datasets to prepare the mineralogical and prospective zone maps of the region and these maps contain the information about the mineral potential zones which were utilized for the perspective of mineral explorations [17, 20, 21, 31, 42, 47, 48, 64–71].

2. Objective of the chapter

This chapter explains the use of SWIR and TIR spectral bands for the demarcation of the different minerals and lithologies present in the region. The importance of integrated datasets from SWIR and TIR-derived outcomes and the utility of the integrated dataset for the demarcation of the various litho-units has also been explained.

3. Study area and geological setup

The coverage of the study area extends between latitude 23°51′35.45″ to 24° 18′34.14″ in the North and longitude 73°28′43.95″ to 73°49′34.24″ in the East and occupies the region in Udaipur and Dungarpur districts of Rajasthan, India (**Figure 1**).

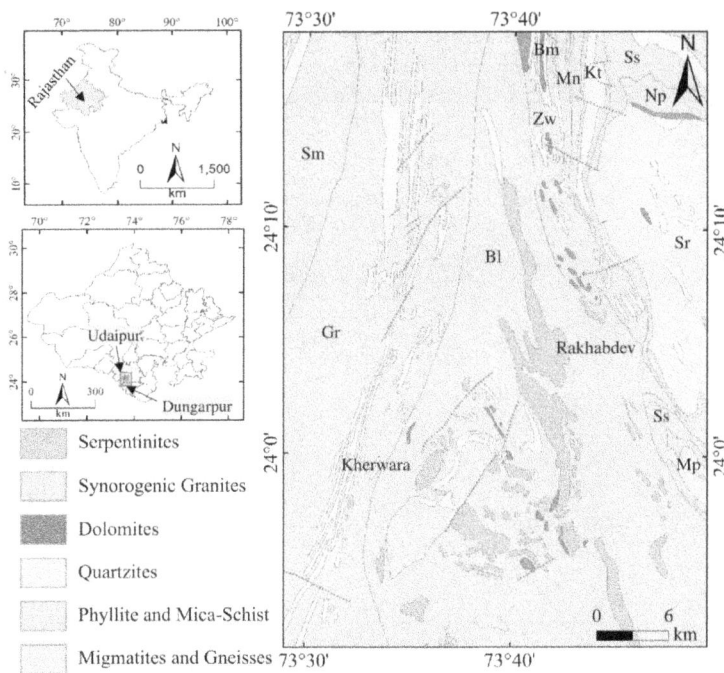

Figure 1.
Location of the study area in inset maps of India and Rajasthan. Lithological map of the study area. Modified after Gupta et al. [72]. Red dashed lines are representing the existing faults. Mp: Mando ki Pal; Sr: Sarada (BGC); Np: Natharia ki Pal; Ss: Sisa Magra; Kt: Kathalia; Mn: Mandli; Bm: Baroi Magra; Bl: Balicha; Zw: Zawar; Gr: Goran; Sm: Samlaji Formations.

Geologically the study area falls in the Udaipur sector exposes various litho-stratigraphic units of the Archean and Palaeoproterozoic age [72–74] (**Figure 1, Table 1**). The basement rocks are the Banded Gneissic Complex (BGC) [73, 75] or the Bhilwara Supergroup (BSG) [72, 74]. They are overlain by the rocks of the Aravalli Supergroup through an erosional unconformity. The Aravalli Supergroup has been categorized into Debari, Udaipur, Bari Lake, Jharol, Dovda, Nathdwara, Lunavada Groups [72]. It has also been subdivided into Lower, Middle, and Upper Aravalli Groups [73].

Majority of the pristine Archean features of the basement rocks have diminished due to tectono-thermal reconstruction of the basement [73]. Basement rocks from Mangalwar Complex are composed of heterogeneous rocks of amphibolite-facies metamorphites [75] or granite-greenstone belt [74]. Gneisses, metabasics, migmatites, and schists constitute the basement while greywacke, chert, marble, dolomite, quartzite, fuchsite quartzite, and mica schist represents the metasediments within the basement [74, 76]. Biotite schist, garnets, and staurolites are present in the Sarara ki Pal inlier [73, 77] and the presence of chlorite and chloritoid represents the retrogression mechanism [78].

The base of the Aravalli Supergroup is having thin bands of quartzites and pebbly oligomictic conglomerate. The continuity of quartzite is interrupted by the ESE-WNW, NE–SW, and ENE-WSW faults. In the majority of the study area phyllites and mica-schists are exposed. Graded bedded greywacke occurs within the phyllite [79]. Poddar & Mathur [80] mentioned the characteristic repetition of graded bedded and slaty phyllite. Different varieties of dolomites are exposed in the Zawar region with gradational contact with greywacke. They are pure to siliceous and massive to gritty nature. Lead-zinc mineralization is confined in the siliceous dolomites [72–75, 77, 81–86]. The lithological and chemical control of the metallogenesis in the region is supporting the concept of the syngenetic origin of lead-zinc sulphides [72]. The Rakhabdev-Dungarpur area consists of ultramafic rocks as linear belts which are serpentinized and are metasomatically altered

Era	Supergroup	Group		Formation
Paleoproterozoic	Aravalli	Synorogenic Granite and Gneiss (intrusion) Rakhabdev Ultramafic Suite (intrusion)		
		Jharol		Samlaji
				Goran
		Udaipur	Tiri Sub-group	Zawar
				Balicha/ Baroi Magra
				Mandli
		Debari		Kathalia
				Sisa Magra
		——Unconformity——		
				Natharia ki Pal
				Gurali/ Basal
		————Unconformity————		
Archean	Banded Gneissic Complex	Mangalwar Complex		Mando ki Pal
				Sarada

Table 1.
Stratigraphic succession of the Aravalli Supergroup from the study area. Modified after Gupta et al. [72].

[73, 87–90]. The ultramafic rocks occur along a prominent lineament named Rakhabdev lineament which passes through the Aravalli fold belt [77, 89]. Thicker ultramafic outcrops are more massive and fractures are developed in an irregular manner [91].

4. Image processing techniques used in this investigation

4.1 Dataset

The present study uses the Advanced Spaceborne Thermal Emission and Reflection Radiometer Level-1 Precision Terrain Corrected Registered At-Sensor Radiance (ASTER L1T) dataset. ASTER sensor carries the VNIR, SWIR, and TIR scanners which have 3 (1–3), 6 (4–9), and 5 (10–14) bands respectively, and its technical specifications mentioned in **Table 2**. The ASTER L1T imagery is already geometrically corrected, georeferenced (WGS-1984) and UTM projected (UTM zone 43 N) [92].

4.2 Methodology

The overall methodology flowchart for the delineation of the various litho-units is depicted in **Figure 2**. The different litho-units were traced out with the help of the Feature-Oriented Principal Component Selection (FPCS) method which uses the various derived outcomes of band ratios, relative band depths, and mineral indices from SWIR and TIR datasets through an integrated approach.

Vegetation and water bodies are present in the region which creates a hindrance in geological mapping therefore, these land features were masked from the derived outcome. Vegetation coverage was calculated using the Normalized Difference Vegetation Index (NDVI) and the values ranging greater than 0.2 were used for the

Granule ID	Sensor-scanner	Band number	Spectral width (μm)	Spatial resolution (mtr)	Radiometric resolution	Valid range
AST_L1T_003042220 03055021_201504280 31510_405 83	ASTER-VNIR		0.520–0.60	15	8-bits	0–255
		2	0.630–0.690	15	8-bits	0–255
		3	0.760–0.860	15	8-bits	0–255
	ASTER-SWIR	4	1.600–1.700	30	8-bits	0–255
		5	2.145–2.185	30	8-bits	0–255
		6	2.185–2.225	30	8-bits	0–255
		7	2.235–2.285	30	8-bits	0–255
		8	2.295–2.365	30	8-bits	0–255
		9	2.360–2.430	30	8-bits	0–255
	ASTER-TIR	10	8.125–8.475	90	12-bits	0–65535
		11	8.475–8.825	90	12-bits	0–65535
		12	8.925–9.275	90	12-bits	0–65535
		13	10.25–10.95	90	12-bits	0–65535
		14	10.95–11.65	90	12-bits	0–65535

Table 2.
Technical specifications of the ASTER L1T dataset. Source: [92].

preparation of the vegetation mask. Water bodies were masked using band 1. The DN values ranging from 0 to 100 were selected to prepare the water mask. Both of the masks were applied on the derived outcomes to eliminate the vegetative lands and water bodies from the mineral and lithological map of the region.

4.2.1 Preprocessing of SWIR and TIR datasets

The ASTER SWIR dataset has the spillover of the energy from band 4 to band 5 and band 9 which is known as crosstalk effects [17, 41, 45, 47, 93, 94]. Crosstalk correction was applied for the removal of effects from the dataset and to enhances the spectral signatures of the minerals/rocks. A semi-empirical atmospheric correction, QUick Atmospheric Correction (QUAC), was applied to retrieve the surface reflection from the sensor radiance [95–100].

The ASTER TIR datasets were converted into the calibrated radiance from the digital number using Eq. (1) [46, 55, 58, 101, 102].

$$L^i_{sen} = cof^i * (DN^i - 1) \tag{1}$$

where:

cof^{10}	cof^{11}	cof^{12}	cof^{13}	cof^{14}
0.006882	0.006780	0.006590	0.005693	0.005224

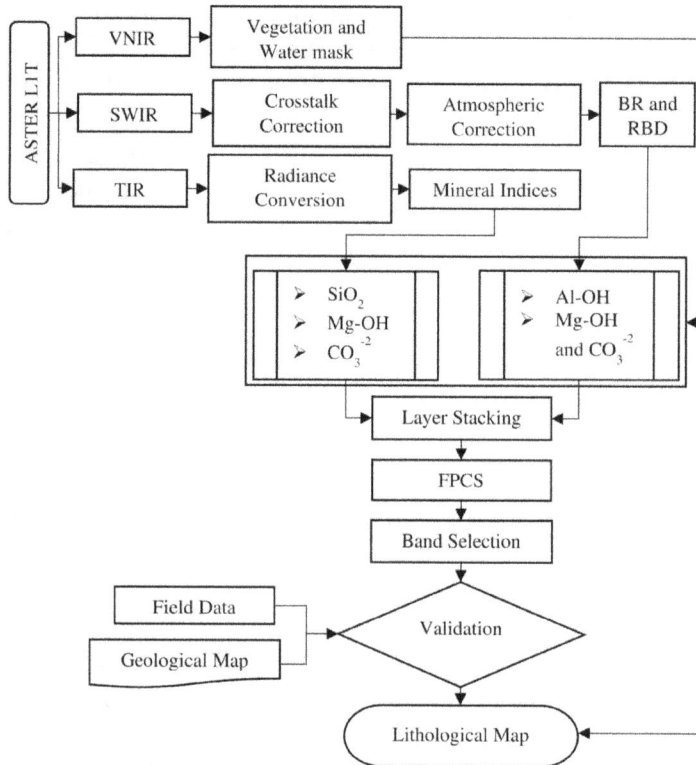

Figure 2.
Overall methodology for the derivation of the lithological map.

4.2.2 Image processing

Band Ratio (BR) is a method in which one band is divided by another band (Eq. (2)) to get the better delineation of the rocks/minerals instead of using a single band and combined with enhancement of spectral properties.

$$BR = B1/B2 \qquad (2)$$

where: BR = Output Band Ratio image; B1 and B2 = Brightness value of selected bands.

Relative Band Depth (RBD) is another technique for mineral mapping in which the position and depth of the mineral spectrum were considered for calculation [103]. The RBD governs better discrimination of minerals than BR because it considers the characteristic absorption features and normalizes the effects generated due to topography and albedo [32, 37, 104]. The bands acquired the shoulder position on absorption spectrum are summed up as (S1 and S2) and divided by the band having minimal absorption value (T; Eq. (3); **Figure 3**).

$$RBD = (S1 + S2)/T \qquad (3)$$

Mineral Indices (MI) is also a mathematical expression derived for mapping of the minerals by using the band math operators with different logics in the TIR wavelength region. The TIR part of the EM spectrum is utilized for the mapping of the feldspars, silicates, carbonates, and ultramafic minerals.

BR and RBD were applied on the atmospherically corrected SWIR datasets and MI was applied on the calibrated radiance TIR datasets (**Table 3**). Al-OH consisting of minerals like mica and clay minerals were delineated with the help of different BRs and RBD6 (**Table 3**). Spectral absorption minima were recorded at band 6 of ASTER at 2.205 µm which highly suitable for the mapping of clay and mica minerals [12, 34, 37, 42, 105]. Similarly, Mg-OH and CO_3^{-2} containing minerals showed the absorption minima at band 8 of ASTER at 2.336 µm which was used in the RBD8 for mapping of carbonates and ultramafics [34, 37, 106, 107]. The SiO_2 containing minerals/rocks showed the emissivity minima at the band 12 of ASTER at 9.075 µm due to vibrational energy along the Si-O bond. The CO_3^{-2} bearing minerals showed the emissivity minima at 11.318 µm which is represented by the band 14. The Mg-OH bearing minerals of ultramafics recorded the emissivity minima at band 13 at 10.657 µm.

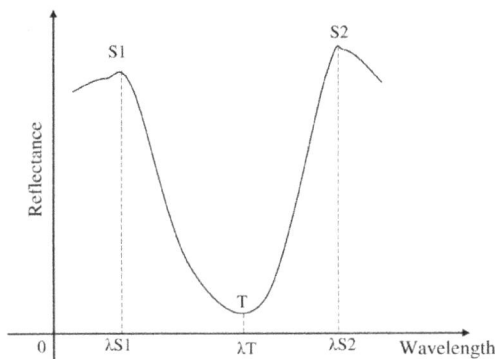

Figure 3.
Artistic sketch for interpretation of the RBD for any mineral.

Principal Component Analysis (PCA) is widely used for the identification and delineation of the litho-units and hydrothermal alteration minerals using the spectral bands generated from different sensors [9, 17, 28, 45, 48, 108]. The PCA uses the statistical mechanism for the transformation of the variables into several linear variables which are not having a correlation with each other, known as Principal Components (PCs). It is implemented on the symmetric matrix which is based on the either correlation matrix or covariance matrix (Eq. (4)). The PCs have the information related to the specific mineral which can be retrieved with the help of loadings of eigenvector (Eq. (5)). The strong eigenvector loadings of the PCs were utilized for the demarcation of the various mineral/groups through its PC indicative bands (Eq. (6)) which showed bright and dark pixels for the specific minerals in the PC image [108]. The present work uses the Feature-Oriented Principal Component Selection (FPCS) on the integrated data derived from BR, RBD, and MI from the ASTER SWIR and TIR data for mapping of the lithological units present in the region. The FPCS was used to achieve the desired goal by the combination of SWIR and TIR-derived outcomes. The phyllite and mica-schist can be marked with the help of BRs and RBD6, carbonates and ultramafics gave a similar tone by the use of RBD8 of SWIR region while MI has the capabilities to distinguish these two minerals/groups. Quartzites were not mapped in the SWIR EM region due to lack of the characteristic absorption band while TIR EM regions have these capabilities. Mineral/rock identification was not possible through the single kind of dataset like only by SWIR only by TIR so, integrated approach was required to delineate all lithological units existed in the study area. Therefore, FPCS was implemented on the basis of covariance matrix of the integrated outcomes of ASTER SWIR and TIR. The derived eigenvector matrix is tabulated in **Table 4**.

$$cov(X, Y) = \frac{1}{n-1}\sum_{i=1}^{n}(Xi - \bar{x})(Yi - \bar{y}) \qquad (4)$$

S. no.	Mineral composition	Indicator minerals	Formula	Absorption band: wavelength (µm)	References
SWIR bands Band Ratio (BR)					
1.	Al-OH	Sericite, smectite, muscovite, and illite	B7/B6	6: 2.205	[106]
2.			B4/B6		[8]
3.		Alunite and kaolinite	B7/B5	5: 2.167	[106]
Relative Band Depth (RBD)					
4.	Al-OH (RBD6)	Sericite, smectite, and illite	(B5 + B8) /B6	6: 2.205	[12, 42]
5.	Mg-OH and CO_3^{-2} (RBD8)	Carbonates and ultramafics	(B6 + B9) /B8	8: 2.336	[37, 106]
TIR bands Mineral Indices (MI)					
6.	Mg-OH	Ultramafics	(B12/B13) × (B14/B13)	13: 10.657	[23]
7.	SiO_2	Silica-rich	(B11/ (B10 + B12)) × (B13/B12)	12: 9.075	[56]
8.	CO_{3-2}	Carbonates	B13/B14	14: 11.318	[54, 101]

Table 3.
BR, RBD, and MI used for the derivation of the mineral maps from ASTER SWIR and TIR bands.

Eigenvectors	MI CO_3^{-2}	MI SiO_2	MI Mg-OH	BR 7/5	RBD6	RBD8	BR 4/6	BR 7/6
Band 1	0.006692	-0.02224	0.056749	-0.18058	-0.50483	0.77256	-0.11678	-0.31352
Band 2	-0.02898	-0.01231	-0.04299	-0.08242	-0.4053	-0.48972	-0.72584	-0.24381
Band 3	-0.69819	-0.33987	-0.62711	-0.0019	0.001132	0.049081	0.036474	0.002323
Band 4	-0.006	-0.01368	0.020563	-0.26053	-0.48361	-0.39945	0.669243	-0.30025
Band 5	0.015369	0.017849	-0.02398	0.807885	-0.49943	-0.00099	0.088098	0.298333
Band 6	-0.0354	-0.00343	0.042673	-0.48543	-0.31084	0.003489	-0.04809	0.813837
Band 7	0.216406	0.736543	-0.6363	-0.05878	-0.02868	0.036048	0.014719	0.000581
Band 8	0.680651	-0.58381	-0.44041	-0.03078	-0.01117	0.007718	0.004385	0.027843

Table 4.
Eigenvector matrix generated from the integrated derived mineral maps for FPCS.

$$A\vec{v} = \lambda\vec{v} \text{ or } \vec{v}(A - \lambda I) = 0 \tag{5}$$

$$y = W' \times x \tag{6}$$

where:

cov(X, Y) = Covariance matrix; X and Y = Variables;

$A\vec{v}$ = eigenvector of matrix A; λ = eigenvalue (scaler value); I = Identity matrix.

y = Final outcome; W' = transpose of scaler data; x = Feature vector.

5. Results

5.1 Band ratios (BRs) and relative band depths (RBDs)

5.1.1 Distribution of Al-OH consisting minerals

Clay (kaolinite, illite, montmorillonite) and mica (sericite and muscovite) minerals consist of the Al-OH in their chemical composition. These minerals especially illite, montmorillonite, muscovite, and sericite exhibited characteristic spectral absorption features at a wavelength of 2.205 µm which is detectable with band 6 (2.185–2.225 µm) of the ASTER sensor. Kaolinite mineral also showed minor absorption at 2.165 µm for that band 5 (2.145–2.185 µm). The Al-OH consisting of minerals were mapped in the quartzites south of Rakhabdev and near the granitic outcrop of Kherwara inlier [89]. Phyllite and mica-schist also depicted higher values for Al-OH containing minerals by using BR 7/6 (**Figure 4A**). The granitoids of the basement, granites, and quartzites have high values for Al-OH by using BR 4/6 (**Figure 4B**). On applying BR 7/5, almost the entire region depicted higher values for kaolinite (Al-OH) which is indicative of a poor interpretation (**Figure 4C**). The RBD (5 + 8)/6 gave a similar kind of result like BR 7/6. It depicted higher values of the Al-OH consisting of rocks/minerals for the quartzites, phyllites, mica-schists, conglomerate, and arkose litho-units (**Figure 4D**). Ultramafics and carbonates (dolomite) have very low values from BRs 7/6, 4/5, and RBD6 due to the absence of Al-OH minerals.

5.1.2 Distribution of Mg-OH and CO_{3-2} consisting minerals

Ultramafics are having the Mg-OH while dolomites are having both Mg-OH and CO_{3-2} constituents in their composition. These minerals have the characteristic absorption feature at 2.33 µm, which occurs at band 8 (2.295–2.365 µm). The RBD (6 + 9)/8 was applied for mapping of the Mg-OH and CO_{3-2} consisting of minerals. Ultramafics depicted extremely high values while dolomites have moderate values (**Figure 5A**). The regions of Zawarmala and Hati Magra are dominated by dolomite exposures but a poor carbonate map as an outcome may be due to the presence of extreme vegetation on the hills.

A lithological map has also been prepared from RBD6, RBD8, and BR4/6 in RGB channels respectively (**Figure 5B**). Basement rocks are depicted as pinkish-blue, phyllite, and mica-schist as reddish color. Dolomite is depicted as dark green while ultramafics are as green color. Quartzites are light pink in color. The resultant map discriminates the various lithologies present in the area and is comparable with the published geological maps.

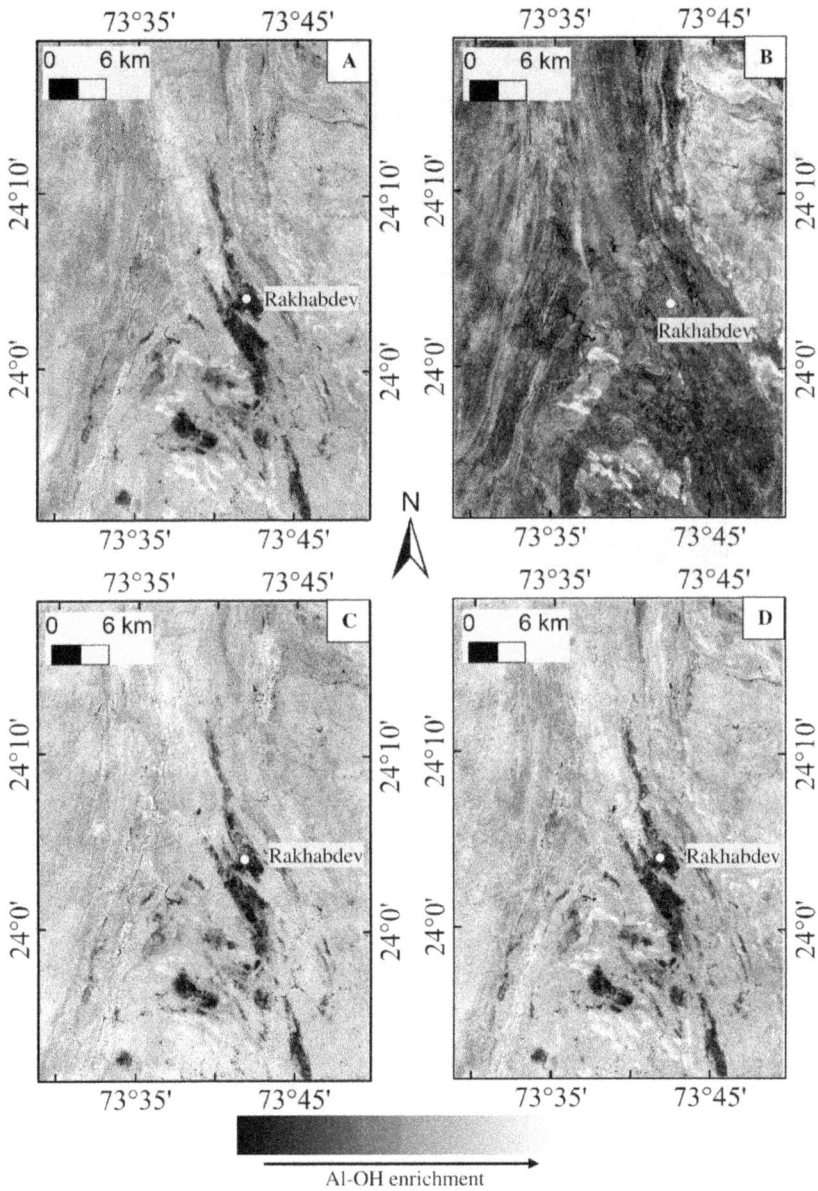

Figure 4.
Results from the SWIR data for Al-OH bearing minerals using the techniques of BRs and RBDs. (A) BR 7/6 for mica and clay minerals. (B) BR 4/6 for clay minerals. (C) BR 7/5 for kaolinite. (D) RBD6 (5 + 8)/6 for clay and mica minerals.

5.2 Mineral indices (MI)

5.2.1 Distribution of mafics/ultramafics

The ultramafic map was developed using the mafic index defined by Guha & Vinod Kumar [23]. The derived map of the mafic index mapped the outcrops of the ultramafics with very higher values near the Rakhabdev region and along other thin

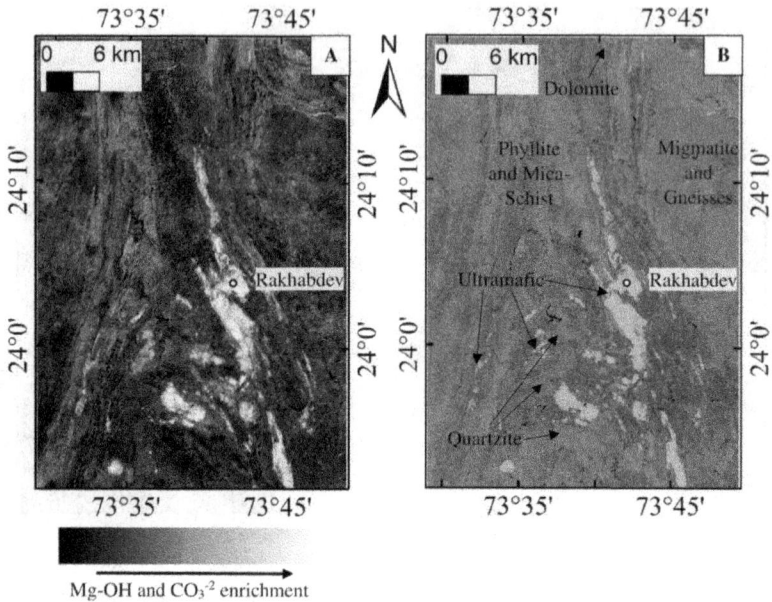

Figure 5.
Results from the SWIR data. (A) RBD8 (6 + 9)/8 for the mapping of ultramafics and carbonates. (B) Lithological map using the RBD6, RBD8, and BR4/6 in the RGB channels.

belts of the ultramafic outcrops south of Rakhabdev and on the west of Kherwara (**Figure 6A**). The dolomites were suppressed and showed their uniqueness to distinguish by ultramafics.

5.2.2 Distribution of SiO_2 consisting minerals

Silica-rich rocks were marked by using the silica index of Rockwell & Hofstra [56] (**Figure 6B**). Quartzites present on the outer periphery of basement rocks were precisely demarcated through the silica index. Silica index also mapped the quartzites present adjacent to the dolomites in the Zawar region. The derived mineral map showed very low values for the ultramafics and dolomites.

5.2.3 Distribution of CO_{3-2} consisting minerals

Ninomiya et al. [54] defined the mathematical expression for the mapping of carbonate rocks and the derived mineral map showed dense noise, consequently, identification of the carbonate outcrops was not precisely obtained (**Figure 6C**). The presence of stripping noise and poor signal at band 14 hinders the demarcation of carbonate outcrops [20, 109, 110]. The ultramafics and quartzitic outcrops were depicting very low values in the carbonate map.

A lithological map was prepared using the silica, mafic, and carbonate indices in RGB channels respectively (**Figure 6D**). Dolomites of the region were marked by bluish-green color but the majority of the region was marked as bluish-green color which is a poor identification for dolomites, quartzites are depicting the maroon color and ultramafics as bright green color. The yellow color at the tips of ultramafics and within the massive outcrops of ultramafics are identified as talc.

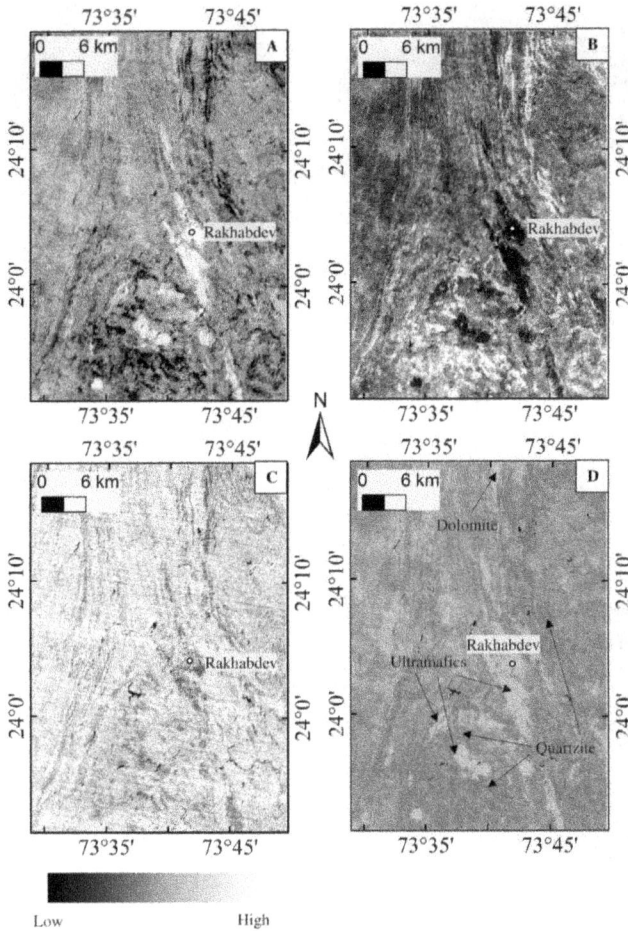

Figure 6.
Results derived from the TIR data. (A) Ultramafic map derived using the Guha and Vinod Kumar index [23]. (B) Silica-rich rocks were demarcated by using the Rockwell and Hofstra index [56]. (C) Carbonate map derived using Ninomiya et al. index [54]. (D) Lithological map prepared using the silica-rich, ultramafics, and carbonate maps in RGB channels.

5.3 Feature-oriented principal component selection (FPCS)

The FPCS technique was implemented to the integrated derived outcomes from BRs, RBDs, and MIs of the ASTER SWIR and TIR bands for discrimination of the different litho-units present in the study area. The generated eigenvector matrix from the FPCS from the integrated derived mineral maps is shown in **Table 4**. The PC1 shows the extreme values for the ultramafics and moderate values for the dolomites. Quartzites showed very low values and represented dark pixels (**Figure 7**). The PC1 shows the combined outcome from the RBD8 and MI Mg-OH because RBD8 highlighted ultramafics and dolomites of the region while MI Mg-OH mapped the ultramafics and suppressed the quartzites of the region. The PC2 showed extremely high values for phyllite and mica-schists and dark pixels for the quartzites of the region. The ultramafics & dolomites of the region are depicting the low values (**Figure 7**). The PC2 showed the combination of BR 4/6, BR 7/5, and RBD6 in which BR 4/6 highlighted the silica-rich rocks. The Al-OH consisting minerals are suppressed in the PC2 while RBD6 highlighted the phyllite and

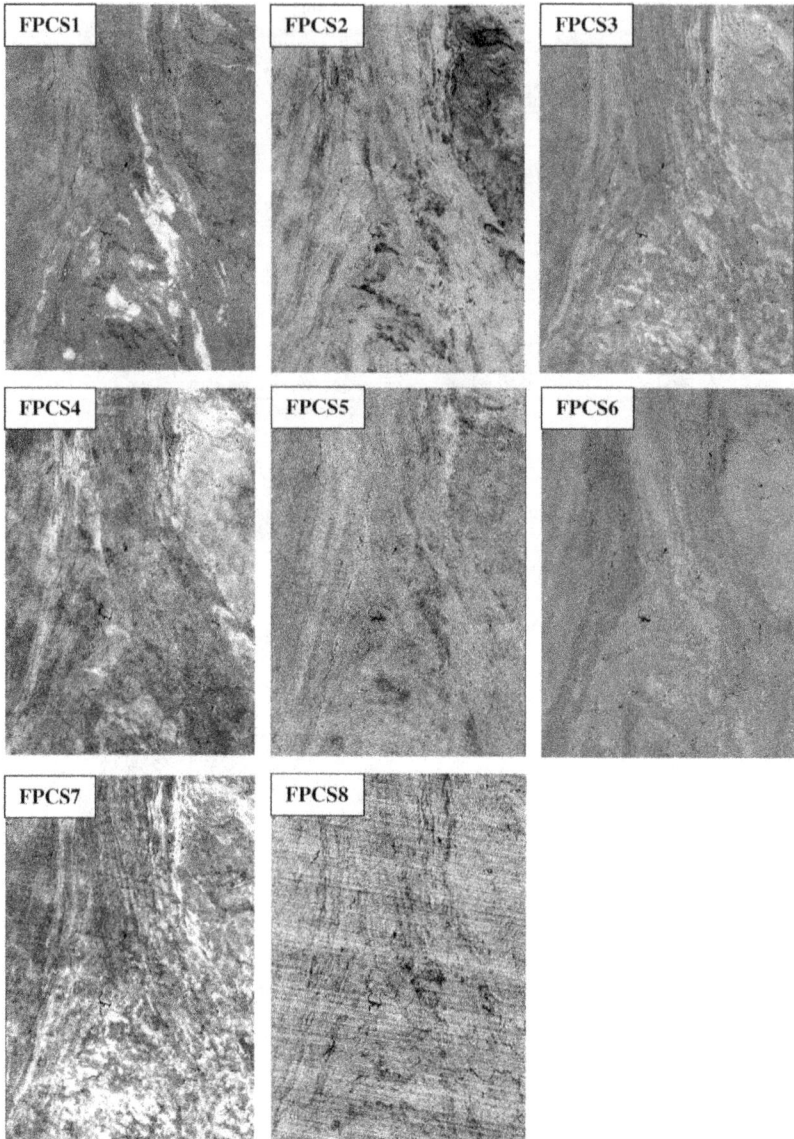

Figure 7.
FPCS components generated using the PCA technique on integrated BRs, RBDs, and MIs from the SWIR and TIR bands.

mica-schists as Al-OH consisting rocks and BR 7/5 also highlighted the Al-OH consisting minerals. The PC3 depicted very high values for quartzites i.e. silica-rich rocks as by MI SiO_2 but on the south of Rakhabdev the distribution of silica-rich rocks is not showing the vague distribution like MI SiO_2, and a clear delineation of quartzites are obtained (**Figure 7**). The PC4 depicted a similar kind of pattern as BR 4/6 but the extremity of the pixel values gets lower down and appearance gets noisy (**Figure 7**). The PC6 depicted higher values for the ultramafics and silica enriched rocks of the granitoids & migmatites from the basement, quartzites, and phyllite, and mica-schist while carbonates are depicted moderate values (**Figure 7**). The PC6 is helpful for the delineation of the litho-units of the region but the band showed an association of the moderate amount of noise. The PC5, PC7, and PC8 are not useful for the discrimination of the geological units due to the presence of a high amount of noise with them (**Figure 7**).

FCC was been prepared using the combinations of bands PC1, PC3, and PC6 respectively for delineation of different litho-units (**Figure 8**). Granitoids and migmatites appeared as greenish-blue colors while quartzites as light green with a mixture of cyan color. Phyllite and mica-schist appeared as dark blue to greenish-brown color. Conglomerate and meta-arkose gave shades of green color. Dolomites appeared as purplish colors and ultramafics as pinkish colors. Light yellow color on the tips of ultramafics and within the ultramafics which showed the presence of talc.

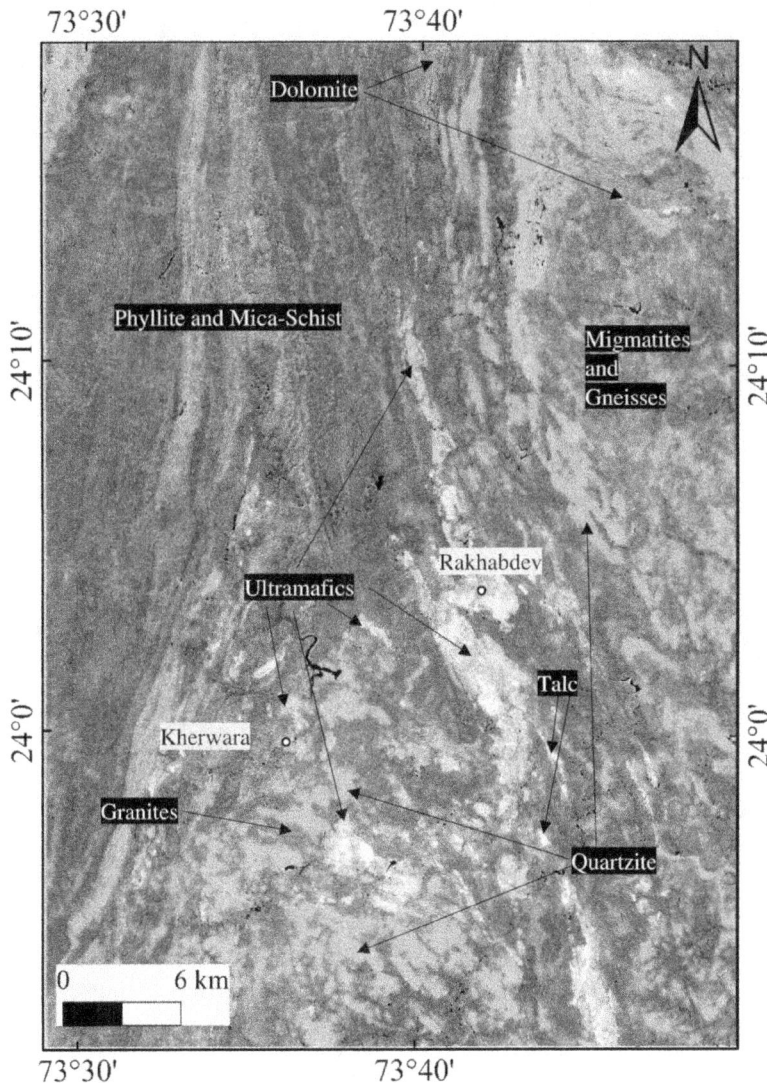

Figure 8.
Lithological map of the study area prepared by the combination of FPCS1, FPCS3, and FPCS6 in the RGB channels respectively.

5.4 Verification of the derived lithological map

To verify the different litho-units of the region various field reconnaissance was conducted and for the estimation of the overall accuracy of the generated lithological map, GPS surveys were carried out. Various traverses were conducted along the major litho-units of the region and some traverses were conducted for the verification of changes observed in the generated lithological map. Field photographs and rock samples were collected for the determination of accurate locations and associated lithology/ies if present at the contact zone. Pink colored granites from the Kherwara Inlier were observed (**Figure 9A**) and deformed quartzites were present with the contact of it (**Figure 9B**). Serpentinites of the Rakhabdev showed the variation all along the belt-like massive to fibrous nature and open mining pits of the serpentinites in the massive variants (**Figure 9C, D, G** and **I**). In the field, outcrops of the different dolomites were observed (**Figure 9E**) associated with the quartzites, metagraywacke, phyllite, and mica-schist (**Figure 9F**). Contact between the phyllite & mica-schist and quartzites was also observed (**Figure 9H**). Talc was also observed in the field and it is mainly in the region of deformation (**Figure 9G**). The isolated patch of the serpentinite near the Parbeela region shows the contact with the granites of Kherwara Inlier. Furthermore, the accuracy assessment was carried out between the generated lithological map and field-collected information

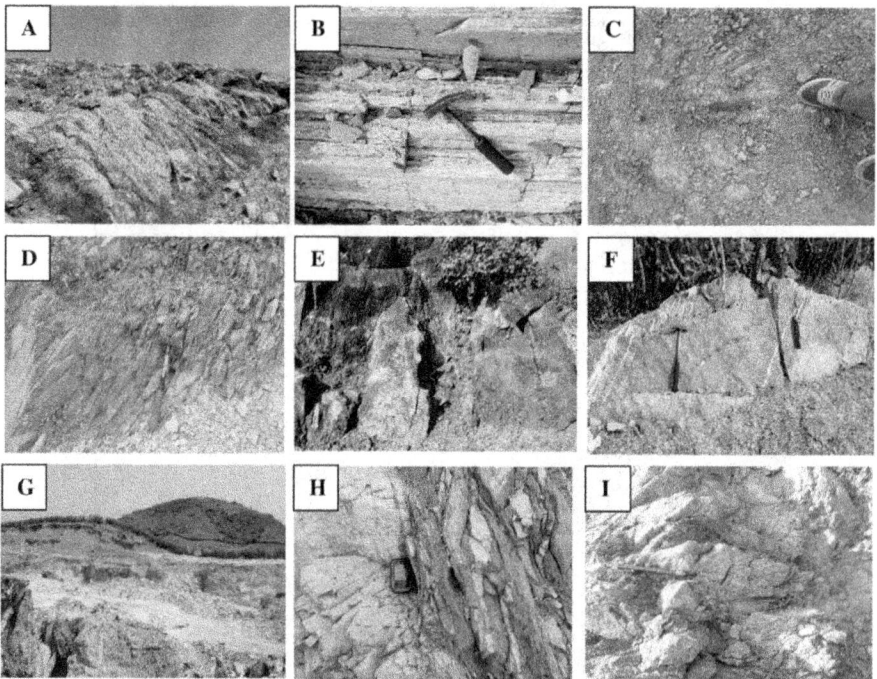

Figure 9.
Field validation of the different litho-units. (A) Granites from the Kherwara inlier. (B) Mullions of quartzites from Kherwara inlier. (C) Talc mineralization along the serpentinites from southeast of Rakhabdev. (D) Massive serpentinites from Rakhabdev. (E) Dolomites from the northeast of Rakhabdev. (F) Massive quartzites from the east of Rakhabdev. (G) Mining activities for talc from the hinges of serpentinites outcrops from south of Rakhabdev. (H) Contact between the quartzites and phyllites from north of Rakhabdev. (I) Serpetninites and its alteration products from south of Rakhabdev.

Class	Dolomite	Serpentinite	Quartzite	Talc	Phyllite	Granite	User's Accuracy
Dolomite	34	1	0	2	0	1	89.47%
Serpentinite	2	16	0	1	0	0	84.21%
Quartzite	0	1	10	0	2	0	76.92%
Talc	2	1	0	14	1	0	77.78%
Phyllite	1	0	2	0	17	1	80.95%
Granite	0	0	0	0	0	13	100.00%
Producer's Accuracy	87.18%	84.21%	83.33%	82.35%	85.00%	86.67%	
			Overall Accuracy				85.25%
			Kappa Coefficient				0.8164

Table 5.
Accuracy assessment of the derived lithological map using the GPS survey collected localities.

using the GPS (**Table 5**). The derived confusion matrix shows the overall accuracy and kappa coefficient as 85.25% and 0.8164 respectively for the lithological map.

6. Discussions

The study area belongs to the Aravalli Orogeny and several deformational histories were recorded [72–75, 77, 81, 84–86, 89, 111–113]. The remote sensing technique is widely used for mineral mapping and lithological mapping around the study area [9, 58, 102, 114]. In the present research, different BR, RBD, & MI and their combinations were generated using the ASTER SWIR and TIR bands for the demarcation of the different litho-units present in the region. The BRs and RBDs were used to derive the Al-OH and Mg-OH & CO_{3-2} consisting minerals [8, 12, 35, 37, 42, 106] like phyllite & mica-schist, carbonates, and ultramafics. And their FCC combinations helped to delineate granitoids, granites, phyllite, mica-schist, quartzites, dolomite, and ultramafics (**Figure 5B**). The MIs were used to derive $SiO2$, Mg-OH, and CO_{3-2} mineral maps of the region [23, 54–56, 58, 101], and quartzites, ultramafics, and carbonates were delineated. The FCC also helped a lot to determine the lithologies of the region but were limited to the quartzites, ultramafics, and carbonates (**Figure 6D**).

FPCS technique was utilized for the delineation of litho-units present in the region on the basis of the combined results derived from the BRs, RBDs, and MIs from the ASTER SWIR and TIR bands. The PC1, PC3, and PC6 have capabilities to discriminate the ultramafics, carbonates, quartzites, phyllite, mica-schist, and granitoids. The prepared FCC using the combinations of these PCs demarcated the granitoids, granites, phyllite, mica-schist, quartzites, conglomerate, meta-arkose, dolomites, ultramafics (**Figure 8**). The integrated approach from the ASTER SWIR and TIR proved its potentials for lithological mapping. The generated confusion matrix showed the overall accuracy as 85.25% and kappa coefficient as 0.8164 in which maximum producer's accuracy (%) was attained by dolomite while user's accuracy (%) by granite. Field validation was performed on the generated lithological map by observing the various litho-units present on the surficial exposures and gathered the location information by the use of a GPS survey. Association of quartzites with dolomites and serpentinites with dolomites was observed in the field. Talc is found with the serpentinites and was produced due to the process of serpentinization and mainly formed at the deformational zones [115].

7. Conclusions

1. Lithological and mineral mapping of the region can be done with the help of various BR, RBD, & MI and by the combinations of these but discrimination between every single litho-unit is not possible with it.

2. The data integration (combination of derived mineral maps from ASTER SWIR and TIR bands) approach played an important role to obtain the desired research goal.

3. PCA is a statistical technique that is utilized for the demarcation of the various litho-units using the original bands but, in this research, FPCS was utilized on the integrated data which shows its capabilities towards discrimination of every single litho-unit.

4. PC1, PC3, and PC6 from the FPCS were utilized for the discrimination of various litho-units in which talc is also identified within ultramafics which is an alteration product of serpentinization.

5. The overall accuracy and kappa coefficient of the generated lithological map are 85.25% and 0.8164 respectively and calculated with the help of GPS surveys.

Acknowledgements

The author is thankful to USGS Earth Explorer to provide the ASTER dataset under the open distribution policy. Thanks to the Heads of the Department of Geology for providing the necessary facilities for conducting the research. Some part of this chapter has been taken from the Ph.D. thesis of the author.

Conflict of interest

The author declares no conflict of interest.

Author details

Ronak Jain[1,2]

1 Department of Remote Sensing, School of Earth Sciences, Banasthali Vidyapith, Banasthali, India

2 Department of Geology, Faculty of Earth Sciences, Mohanlal Sukhadia University, Udaipur, India

*Address all correspondence to: jainronak75@yahoo.in

IntechOpen

References

[1] Pandey SN. Principles and Applications of Photogeology. 1st ed. United States: John Wiley and Sons Ltd; 1984. 1-382 p.

[2] Guha PK. Remote Sensing for the Beginner. New Delhi, India: East-West Press Ltd.; 2003. 129 p.

[3] Gupta RP. Introduction. In: Remote Sensing Geology. 2nd ed. New York: Springer- Verlag Berlin Heidelberg GmbH; 2003. p. 1–16.

[4] Jain R, Kumar A, Sharma RU. Study of Mineral Mapping Techniques using Airborne Hyperspectral Data: Exploring the potential of AVIRIS-NG for Mineral Identification. Germany: Lap Lambert Academic Publishing; 2018. 1-72 p.

[5] Jain R, Sharma RU. Airborne hyperspectral data for mineral mapping in Southeastern Rajasthan, India. Int J Appl Earth Obs Geoinf. 2019;81:137–145.

[6] Jensen JR. Introductory Digital Image Processing: A Remote Sensing Perspective. 4th ed. Pearson Education, Inc; 2015. 623 p.

[7] Jain R, Kumar A, Sharma RU. Study of Mineral Mapping Techniques: A case study in Southeastern Rajasthan. In: Proceedings of 38th Asian Conference on Remote Sensing- Space Applications: Touching Human Lives, ACRS. New Delhi, India: Curran Associates, Inc., New York; 2017. p. 2799–807.

[8] Sabins FF. Remote Sensing: Principles and Interpretation. 3rd ed. United States of America: Waveland Press, Inc.; 2007. 1-485 p.

[9] Kumar C, Shetty A, Raval S, Sharma R, Ray PKC. Lithological Discrimination and Mapping using ASTER SWIR Data in the Udaipur area of Rajasthan, India. Procedia Earth Planet Sci. 2015;11:180–188.

[10] Ninomiya Y. Toward Lithological Mapping of Arabian Peninsula Using ASTER Multispectral Thermal Infrared Data. In: El-Askary H, Lee S, Heggy E, Pradhan B, editors. Advances in Remote Sensing and Geo Informatics Applications CAJG 2018 Advances in Science, Technology & Innovation (IEREK Interdisciplinary Series for Sustainable Development). Springer, Cham; 2019. p. 181–184.

[11] Ninomiya Y, Fu B. Wide area lithologic mapping with ASTER thermal infrared data: Case studies for the regions in/around the Pamir Mountains and the Tarim basin. IOP Conf Ser Earth Environ Sci. 2017;74(1):1–4.

[12] Rowan LC, Mars JC, Simpson CJ. Lithologic mapping of the Mordor, NT, Australia ultramafic complex by using the Advanced Spaceborne Thermal Emission and Reflection Radiometer (ASTER). Remote Sens Environ. 2005; 99(1–2):105–126.

[13] Vincheh ZH, Arfania R. Lithological Mapping from OLI and ASTER Multispectral Data Using Matched Filtering and Spectral Analogues Techniques in the Pasab-e-Bala Area, Central Iran. Open J Geol [Internet]. 2017;7:1494–508. Available from: http://www.scirp.org/journal/ojg

[14] Ozkan M, Celik O, Ozyavas A. Lithological discrimination of accretionary complex (Sivas, northern Turkey) using novel hybrid color composites and field data. J African Earth Sci. 2018;138:75–85.

[15] Madani AA, Emam AA. SWIR ASTER band ratios for lithological mapping and mineral exploration: a case study from El Hudi area, southeastern desert, Egypt. Arab J Geosci. 2011;4(1–2):45–52.

[16] Askari G, Pour A, Pradhan B, Sarfi M, Nazemnejad F. Band Ratios

Matrix Transformation (BRMT): A Sedimentary Lithology Mapping Approach Using ASTER Satellite Sensor. Sensors. 2018;18(10):3213.

[17] Pour AB, Hashim M. The application of ASTER remote sensing data to porphyry copper and epithermal gold deposits. Ore Geol Rev. 2012;44: 1–9.

[18] Pal SK, Majumdar TJ, Bhattacharya AK, Bhattacharyya R. Utilization of Landsat ETM+ data for mineral-occurrences mapping over Dalma and Dhanjori, Jharkhand, India: an Advanced Spectral Analysis approach. Int J Remote Sens. 2011;32 (14):4023–4040.

[19] Zhang X, Pazner M, Duke N. Lithologic and mineral information extraction for gold exploration using ASTER data in the south Chocolate Mountains (California). ISPRS J Photogramm Remote Sens. 2007;62(4): 271–282.

[20] Guha A, Yamaguchi Y, Chatterjee S, Rani K, Vinod Kumar K. Emittance Spectroscopy and Broadband Thermal Remote Sensing Applied to Phosphorite and Its Utility in Geoexploration: A Study in the Parts of Rajasthan, India. Remote Sens. 2019;11(9):1003.

[21] Rani K, Guha A, Mondal S, Pal SK, Vinod Kumar K. ASTER multispectral bands, ground magnetic data, ground spectroscopy and space-based EIGEN6C4 gravity data model for identifying potential zones for gold sulphide mineralization in Bhukia, Rajasthan, India. J Appl Geophys. 2019; 160:28–46.

[22] Guha A, Singh VK, Parveen R, Vinod Kumar K, Jeyaseelan, A. T. Dhanamjaya Rao EN. Analysis of ASTER data for mapping bauxite rich pockets within high altitude lateritic bauxite, Jharkhand, India. Int J Appl Earth Obs Geoinf. 2013;21:184–194.

[23] Guha A, Vinod Kumar K. New ASTER derived thermal indices to delineate mineralogy of different granitoids of an Archaean Craton and analysis of their potentials with reference to Ninomiya's indices for delineating quartz and mafic minerals of granitoids—An analysis in Dharwar Craton. Ore Geol Rev. 2016;74:76–87.

[24] Guha A, Vinod Kumar K, Dhananjaya Rao EN, Parveen R. An image processing approach for converging ASTER-derived spectral maps for mapping Kolhan limestone, Jharkhand, India. Curr Sci. 2014;106(1): 40–49.

[25] Yajima T, Yamamoto K, Yamamoto K, Hayashi T. Identification of hydrothermal alteration zones for exploration of porphyry copper deposits using ASTER data. J Remote Sens Soc Japan. 2007;27(2):117–128.

[26] Pour AB, Hashim M. Hydrothermal alteration mapping from Landsat-8 data, Sar Cheshmeh copper mining district, south-eastern Islamic Republic of Iran. J Taibah Univ Sci [Internet]. 2015;9(2):155–166. Available from: h ttp://dx.doi.org/10.1016/j.jtusc i.2014.11.008

[27] Pour AB, Hashim M, Hong JK, Park Y. Lithological and alteration mineral mapping in poorly exposed lithologies using Landsat-8 and ASTER satellite data: North-eastern Graham Land, Antarctic Peninsula. Ore Geol Rev. 2019;108:112–133.

[28] Sekandari M, Masoumi I, Pour AB, Muslim AM, Hossain MS, Misra A. ASTER and WorldView-3 satellite data for mapping lithology and alteration minerals associated with Pb-Zn mineralization. Geocarto Int. 2020;1–31.

[29] Abdelsalam MG, Stern RJ, Berhane WG. Mapping gossans in arid regions with Landsat TM and SIR-C images: The Beddaho Alteration Zone in

northern Eritrea. J African Earth Sci. 2000;30(4):903–916.

[30] Khaleghi M, Ranjbar H. Alteration Mapping for Exploration of Porphyry Copper Mineralization in the Sarduiyeh Area, Kerman Province, Iran, Using ASTER SWIR Data. Aust J Basic Appl Sci. 2011;5(8):61–69.

[31] Honarmand M. Application of Airborne Geophysical and ASTER Data for Hydrothermal Alteration Mapping in the Sar-Kuh Porphyry Copper Area, Kerman Province, Iran. Open J Geol [Internet]. 2016;6:1257–1268. Available from: http://www.scirp.org/journal/ojg

[32] Yan J, Zhou K, Liu D, Wang J, Wang L, Liu H. Alteration information extraction using improved relative absorption band-depth images, from HJ-1A HSI data: a case study in Xinjiang Hatu gold ore district. Int J Remote Sens. 2014;35(18):6728–6741.

[33] Honarmand M, Ranjbar H, Shahabpour J. Application of Principal Component Analysis and Spectral Angle Mapper in the Mapping of Hydrothermal Alteration in the Jebal–Barez Area, Southeastern Iran. Resour Geol. 2012;62(2):119–139.

[34] Zhang T, Yi G, Li H, Wang Z, Tang J, Zhong K, et al. Integrating Data of ASTER and Landsat-8 OLI (AO) for Hydrothermal Alteration Mineral Mapping in Duolong Porphyry Cu-Au Deposit, Tibetan Plateau, China. Remote Sens. 2016;8(11):890.

[35] Parashar C. Mapping of Alteration mineral zones by combining techniques of Remote Sensing and Spectroscopy in the parts of SE-Rajasthan. Andra University, Visakhapatnam; 2015.

[36] Moghtaderi A, Moore F, Mohammadzadeh A. The application of advanced space-borne thermal emission and reflection (ASTER) radiometer data in the detection of alteration in the

Chadormalu paleocrater, Bafq region, Central Iran. J Asian Earth Sci. 2007;30 (2):238–252.

[37] Sengar VK, Venkatesh AS, Champati Ray PK, Sahoo PR, Khan I, Chattoraj SL. Spaceborne mapping of hydrothermal alteration zones associated with the Mundiyawas-Khera copper deposit, Rajasthan, India, using SWIR bands of ASTER: Implications for exploration targeting. Ore Geol Rev. 2020;118:103327.

[38] Tommaso ID, Rubinstein N. Hydrothermal alteration mapping using ASTER data in the Infiernillo porphyry deposit, Argentina. Ore Geol Rev. 2007; 32(1–2):275–290.

[39] Lampinen HM, Laukamp C, Occhipinti SA, Metelka V, Spinks SC. Delineating Alteration Footprints from Field and ASTER SWIR Spectra, Geochemistry, and Gamma-Ray Spectrometry above Regolith-Covered Base Metal Deposits - An Example from Abra, Western Australia. Econ Geol. 2017;112(8):1977–2003.

[40] Pour AB, Hashim M. Spectral transformation of ASTER data and the discrimination of hydrothermal alteration minerals in a semi-arid region, SE Iran. Int J Phys Sci. 2011;6(8):2037–2059.

[41] Pour AB, Hashim M. Identification of hydrothermal alteration minerals for exploring of porphyry copper deposit using ASTER data, SE Iran. J Asian Earth Sci. 2011;42(6):1309–1323.

[42] Van der Meer FD, Van der Werff HMA, Van Ruitenbeek FJA, Hecker CA, Bakker WH, Noomen MF, et al. Multi- and hyperspectral geologic remote sensing: A review. Int J Appl Earth Obs Geoinf [Internet]. 2012;14(1): 112–128. Available from: http://dx.doi. org/10.1016/j.jag.2011.08.002

[43] Amri K, Mahdjoub Y, Guergour L. Use of Landsat 7 ETM+ for lithological

and structural mapping of Wadi Afara Heouine area (Tahifet–Central Hoggar, Algeria). Arab J Geosci. 2011;4(7–8): 1273–1287.

[44] Al-Nahmi F, Saddiqi O, Hilali A, Rhinane H, Baidder L, El Arabi H, et al. Application of remote sensing in geological mapping, case study Al Maghrabah area – Hajjah region, Yemen. In: ISPRS Annals of the Photogrammetry, Remote Sensing and Spatial Information Sciences. Safranbolu, Karabuk, Turkey; 2017. p. 63–71.

[45] Sheikhrahimi A, Pour AB, Pradhan B, Zoheir B. Mapping hydrothermal alteration zones and lineaments associated with orogenic gold mineralization using ASTER data: A case study from the Sanandaj-Sirjan Zone, Iran. Adv Sp Res. 2019;63(10): 3315–3332.

[46] Pour AB, Park Y, Crispini L, Läufer A, Hong JK, Park TS, et al. Mapping Listvenite Occurrences in the Damage Zones of Northern Victoria Land, Antarctica Using ASTER Satellite Remote Sensing Data. Remote Sens. 2019;11(12):1408.

[47] Chattoraj SL, Prasad G, Sharma RU, Champati Ray PK, Van der Meer FD, Guha A, et al. Integration of remote sensing, gravity and geochemical data for exploration of Cu-mineralization in Alwar basin, Rajasthan, India. Int J Appl Earth Obs Geoinf. 2020;91:102162 (1-12).

[48] Sekandari M, Masoumi I, Pour AB, Muslim AM, Rahmani O, Hashim M, et al. Application of Landsat-8, Sentinel-2, ASTER and WorldView-3 Spectral Imagery for Exploration of Carbonate-Hosted Pb-Zn Deposits in the Central Iranian Terrane (CIT). Remote Sens. 2020;12(8):1239 (1-33).

[49] Van der Meer FD. Imaging spectrometry for geological remote sensing. Netherlands J Geosci. 1998;77 (2):137–151.

[50] Thompson AJB, Hauff PL, Robitaille AJ. Alteration mapping in exploration; application of short-wave infrared SWIR spectroscopy. SEG Newsl. 1999;39:16–27.

[51] Thompson AJB, Hauff PL, Robitaille AJ. Alteration Mapping in Exploration: Application of Short-Wave Infrared (SWIR) Spectroscopy. In: Bedell R, Crósta AP, Grunsky E, editors. Remote Sensing and Spectral Geology. McLean, Va: Society of Economic Geologists; 2009.

[52] Jain R, Sharma RU. Mapping of Mineral Zones using the Spectral Feature Fitting Method in Jahazpur belt, Rajasthan, India. Int Res J Eng Technol. 2018;5(1):562–567.

[53] Chang CW, Laird DA, Mausbach MJ, Hurburgh CR. Near-infrared reflectance spectroscopy - Principal components regression analyses of soil properties. Soil Sci Soc Am J. 2001;65(2):480–490.

[54] Ninomiya Y, Fu B, Cudahy TJ. Detecting lithology with advanced spaceborne thermal emission and reflection radiometer (ASTER) multispectral thermal infrared "radiance-at-sensor" data. Remote Sens Environ. 2005;99(1–2):127–139.

[55] Ninomiya Y, Fu B. Regional Lithological Mapping Using ASTER-TIR Data: Case Study for the Tibetan Plateau and the Surrounding Area. Geosciences. 2016;6:39.

[56] Rockwell BW, Hofstra AF. Identification of quartz and carbonate minerals across northern Nevada using ASTER thermal infrared emissivity data- implications for geologic mapping and mineral resource investigations in well-studied and frontier areas. Geosphere. 2008;4(1):218–246.

[57] Rani K, Guha A, Pal SK, Vinod Kumar K. Comparative analysis of potentials of ASTER thermal infrared band derived emissivity composite, radiance composite and emissivity-temperature composite in geological mapping of Proterozoic rocks in parts Banswara, Rajasthan. J Indian Soc Remote Sens. 2018;46(5):771–782.

[58] Jain R, Bhu H, Purohit R. Application of Thermal Remote Sensing technique for mapping of ultramafic, carbonate and siliceous rocks using ASTER data in Southern Rajasthan, India. Curr Sci. 2020;119(6):954–961.

[59] Rajendran S, Hersi OS, Al-Harthy A, Al-Wardi M, El-Ghali MA, Al-Abri AH. Capability of Advanced Spaceborne Thermal Emission and Reflection Radiometer (ASTER) on discrimination of carbonates and associated rocks and mineral identification of eastern mountain region (Saih Hatat window) of Sultanate of Oman. Carbonates and Evaporites. 2011;26(4):351–364.

[60] Rajendran S, Nasir S. ASTER spectral sensitivity of carbonate rocks — study in Sultanate of Oman. Adv Sp Res. 2014;53(4):656–673.

[61] Pour AB, Park TYS, Park Y, Hong JK, Zoheir B, Pradhan B, et al. Application of Multi-Sensor Satellite Data for Exploration of Zn–Pb Sulfide Mineralization in the Franklinian Basin, North Greenland. Remote Sens. 2018;10(8):1186.

[62] Yao K, Pradhan B, Idrees MO. Identification of Rocks and Their Quartz Content in Gua Musang Goldfield Using Advanced Spaceborne Thermal Emission and Reflection Radiometer Imagery. Jung HS, editor. J Sensors. 2017;2017:6794095.

[63] El Janati M, Soulaimani A, Admou H, Youbi M, Hafid A, Hefferan K. Application of ASTER remote sensing data to geological mapping of basement domains in arid regions: a case study from the Central Anti-Atlas, Iguerda inlier, Morocco. Arab J Geosci. 2014;7(6):2407–2422.

[64] Guha A, Chatterjee S, Oommen T, Vinod Kumar K, Roy SK. Synergistic use of ASTER, L-band ALOS PALSAR, and Hyperspectral AVIRIS-NG data for exploration of lode type gold deposit - A study in Hutti Maski Schist Belt, India. Ore Geol Rev. 2021;128:103818.

[65] Guha A, Chattoraj SL, Chatterjee S, Vinod Kumar K, Rao PVN, Bhaumik AK. Reflectance spectroscopy-guided broadband spectral derivative approach to detect glauconite-rich zones in fossiliferous limestone, Kachchh region, Gujarat, India. Ore Geol Rev. 2020;127:103825.

[66] Pour AB, Hashim M. ASTER, ALI and Hyperion sensors data for lithological mapping and ore minerals exploration. Springer Plus. 2014;3:130.

[67] Asadzadeh S, de Souza Filho CR. A review on spectral processing methods for geological remote sensing. International Journal of Applied Earth Observation and Geoinformation. 2016.

[68] Abbaszadeh M, Hezarkhani A. Enhancement of hydrothermal alteration zones using the spectral feature fitting method in Rabor area, Kerman, Iran. Arab J Geosci. 2013;6:1957–1964.

[69] Goodarzi Mehr S, Ahadnejad V, Abbaspour RA, Hamzeh M. Using the mixture-tuned matched filtering method for lithological mapping with Landsat TM5 images. Int J Remote Sens. 2013;34(24):8803–8816.

[70] Zadeh MH, Tangestani MH, Roldan FV, Yusta I. Mineral exploration and alteration zone mapping using mixture tuned matched filtering approach on ASTER data at the central part of dehaj-sarduiyeh copper belt, SE

Iran. IEEE J Sel Top Appl Earth Obs Remote Sens. 2014;7(1):284–289.

[71] Porwal A, Carranza EJM, Hale M. A Hybrid Neuro-Fuzzy Model for Mineral Potential Mapping. Math Geol. 2004;36 (7):803–826.

[72] Gupta SN, Arora YK, Mathur RK, Iqballuddin, Prasad B, Sahai TN, et al. The Precambrian Geology of the Aravalli region, Southern Rajasthan & North-eastern Gujarat. Mem Geol Surv India. 1997;123:1–262.

[73] Roy AB, Jakhar SR. Geology of Rajasthan (Northwest India): Precambrian to Recent. Jodhpur: Scientific Publishers (India); 2002. 421 p.

[74] GSI. Geology and Mineral Resources of Rajasthan. 3rd ed. Kolkata, India: Geological Survey of India; 2011. 130pp.

[75] Sinha-Roy S, Malhotra G, Mohanty M. Geology of Rajasthan. Bangalore: Geological Society of India; 1998. 1-275 p.

[76] Gupta SN, Arora YK, Mathur RK, Iqballuddin, Prasad B, Sahai TN, et al. Lithostratigraphic map of Aravalli region, southern Rajasthan & northern Gujarat. Hyderabad: Geological Survey of India; 1980.

[77] Heron AM. Geology of the Central Rajasthan. Mem Geol Surv India. 1953; 79:1–389.

[78] Bhattacharyya S, Dutt K, Sarkar SS. Detailed Study of Mangalwar Complex. Abstracts of Progress reports: 1993-93. Rec Geol Surv India. 1995;127(7):1–3.

[79] Mathur RK. Systematic geological mapping in parts of Udaipur district, Rajasthan. Calcutta, India; 1964.

[80] Poddar BC, Mathur RK. A note on the repetitive sequence of greywacke-slate–phyllite in the Aravalli System around Udaipur, Rajasthan. Bull Geol Soc India. 1965;2(2):192–194.

[81] Straczek JA, Srikantan B. The Geology of the Zawar Lead–Zinc Area, Rajasthan, India. Mem Geol Surv India. 1966;92:1–85.

[82] Mookherjee A. Geology of the Zawar Lead-Zinc Mine, Rajasthan, India. Econ Geol. 1964;59(4):656–677.

[83] Poddar BC. Lead-Zinc mineralization in the Zawar Belt, India - Discussion. Econ Geol. 1965;60(3):636–638.

[84] Singh NN. Tectonic and stratigraphic framework of the lead-zinc sulphide mineralisation at Zawarmala, District Udaipur, Rajasthan. J Geol Soc India. 1988;31(6):546–564.

[85] Roy AB. Geometry and evolution of superposed folding in the Zawar lead-zinc mineralised belt, Rajasthan. Proc Indian Acad Sci (Earth Planet Sci. 1995; 104(3):349–71.

[86] Roy AB, Jain AK. Polyphase deformation in the Pb-Zn bearing Precambrian rocks of Zawarmala, Udaipur district, southern Rajasthan. Q J Geol Min Metall Soc India. 1974;46:81–86.

[87] Bhu H, Sarkar A, Purohit R, Banerjee A. Characterization of fluid involved in ultramafic rocks along the Rakhabdev Lineament from southern Rajasthan, northwest India. Curr Sci. 2006;91(9):1251–1256.

[88] Purohit R, Bhu H, Sarkar A, Ram J. Evolution of the ultramafic rocks of the Rakhabdev and Jharol belts in southeastern Rajasthan, India: New evidences from imagery mapping, petro-minerological and OH stable isotope studies. J Geol Soc India. 2015;85 (3):331–338.

[89] Ram J. Tectonism along the Rakhabdev Lineament as Exemplified by Structural and Crustal Deformation

Studies. Mohanlal Sukhadia University, Udaipur; 2014.

[90] Sarkar DP, Ando JI, Das K, Chattopadhyay A, Ghosh G, Shimizu K, et al. Serpentinite enigma of the Rakhabdev lineament in western India: Origin, deformation characterization and tectonic implications. J Mineral Petrol Sci. 2020;115(2):216–226.

[91] Chattopadhyay A, Gangopadhyay S. Petrological Studies of the Ultramafics rocks of the Rajasthan. Geol Surv India, Spec Publ. 1984;12:17–24.

[92] USGS. AST_L1T v003 [Internet]. LPDAAC. 2019 [cited 2020 Jan 13]. Available from: https://lpdaac.usgs.gov/products/ast_l1tv003/

[93] Iwasaki A, Tonooka H. Validation of a Crosstalk Correction Algorithm for ASTER/SWIR. IEEE Trans Geosci Remote Sens. 2005;43(12):2747–2751.

[94] Mars JC, Rowan LC. Spectral assessment of new ASTER SWIR surface reflectance data products for spectroscopic mapping of rocks and minerals. Remote Sens Environ. 2010; 114(9):2011–2025.

[95] Bernstein LS, Jin X, Gregor B, Adler-Golden SM. The Quick Atmospheric Correction (QUAC) Code: Algorithm Description and Recent Upgrades. SPIE Opt Eng. 2012;51(11):111719.

[96] Bernstein LS, Adler-Golden SM, Sundberg RL, Levine RY, Perkins TC, Berk A, et al. Validation of the QUick atmospheric correction (QUAC) algorithm for VNIR-SWIR multi- and hyperspectral imagery. In: Shen SS, Lewis PE, editors. Proceedings of SPIE 5806, Algorithms and Technologies for Multispectral, Hyperspectral, and Ultraspectral Imagery XI. SPIE; 2005. p. 668–678.

[97] Bernstein LS, Adler-Golden SM, Sundberg RL, Levine RY, Perkins TC,

Berk A, et al. A New Method for Atmospheric Correction and Aerosol Optical Property Retrieval for VIS-SWIR Multi- and Hyperspectral Imaging Sensors: Quick Atmospheric Correction. In: Green RO, editor. Proceedings of 13th JPL Airborne Earth Science Workshop. Pasadena, CA: Jet Propulsion Laboratory, California Institute of Technology; 2004. p. 9–20.

[98] Saini V, Tiwari RK, Gupta RP. Comparison of FLAASH and QUAC atmospheric correction methods for Resourcesat-2 LISS-IV data. In: Proceedings of SPIE 9881, In Earth observing missions and sensors: Development, implementation, and characterization IV. New Delhi, India: SPIE; 2016.

[99] Zhu S, Lei B, Wu Y. Retrieval of Hyperspectral Surface Reflectance Based on Machine Learning. Remote Sens. 2018;10(2):323.

[100] Lentilucci EJ, Adler-Golden SM. Atmospheric Compensation of Hyperspectral Data: An Overview and Review of In-Scene and Physics-Based Approaches. IEEE Geosci Remote Sens Mag. 2019;7(2):31–50.

[101] Ninomiya Y, Fu B. Thermal infrared multispectral remote sensing of lithology and mineralogy based on spectral properties of materials. Ore Geol Rev. 2019;108:54–72.

[102] Jain R. Geological Studies using the Imaging Spectroscopy and Polarimetric Synthetic Aperture Radar (SAR) data in Zawar (Distt: Udaipur). Mohanlal Sukhadia University, Udaipur; 2021.

[103] Van der Meer FD. Analysis of spectral absorption features in Hyperspectral Imagery. Int J Appl Earth Obs Geoinf. 2004;5(1):55–68.

[104] Wang JN, Zheng LF. The Spectral Absorption Identification Model and Mineral Mapping by Imaging

Spectrometer Data. Remote Sens Environ China. 1996;1:20–31.

[105] Mars JC, Rowan LC. Regional mapping of phyllic- and argillic-altered rocks in the Zagros magmatic arc, Iran, using Advanced Spaceborne Thermal Emission and Reflection Radiometer (ASTER) data and logical operator algorithms. Geosphere. 2006;2(3): 161–186.

[106] Hewson RD, Cudahy TJ, Mizuhiko S, Ueda K, Mauger AJ. Seamless geological map generation using ASTER in the Broken Hill-Curnamona province of Australia. Remote Sens Environ. 2005;99(1–2): 159–172.

[107] Cudahy TJ, Hewson R, Buchanan A, Maruyama Y, Mauger Creasey J, Veridan. A generation of geological and regolith maps derived from multispectral VNIR-SWIR-TIR ASTER satellite data. In: Proceedings of the Fourteenth International Conference on Applied Geologic Remote Sensing. USA; 2000. p. 159.

[108] Crosta AP, De Souza Filho CR, Azevedo F, Brodie C. Targeting key alteration minerals in epithermal deposits in Patagonia, Argentina, using ASTER imagery and principal component analysis. Int J Remote Sens. 2003;24(21):4233–4240.

[109] Yajima T, Yamaguchi Y. Geological mapping of the Francistown area in northeastern Botswana by surface temperature and spectral emissivity information derived from Advanced Spaceborne Thermal Emission and Reflection Radiometer (ASTER) thermal infrared data. Ore Geol Rev. 2013;53:134–144.

[110] Son YS, Kang MK, Yoon WJ. Lithological and mineralogical survey of the Oyu Tolgoi region, Southeastern Gobi, Mongolia using ASTER reflectance and emissivity data. Int J

Appl Earth Obs Geoinf. 2014;26: 205–216.

[111] Roy AB, Bejarniya BR. A tectonic model for the early Proterozoic Aravalli (Supergroup) rock from north of Udaipur, Rajasthan. In: Sychanthavong SPH, editor. Crustal Evolution and Orogeny. New Delhi, India: Oxford and IBH Publishing Co. Pvt. Ltd.; 1990. p. 249–273.

[112] Verma PK, Greiling RO. Tectonic evolution of the Aravalli orogen (NW India): an inverted Proterozoic rift basin? Geol Rundsch. 1995;84(4):683–696.

[113] Roy AB. Stratigraphic and tectonic framework of the Aravalli Mountain Range. In : A.B. Roy (Eds.) Precambrians of Aravalli Mountain, Rajasthan, India. Mem Geol Soc India. 1988;7:3–31.

[114] Salaj SS, Prabhakaran, Upadhyay R, Srivastav SK. Mineral abundance mapping using hyperion dataset in Udaipur, India. Geospatial World [Internet]. 2012; Available from: https://www.geospatialworld.net/artic le/mineral-abundance-mapping-using-h yperion-dataset-in-udaipur-india/

[115] Kumar H, Rajawat AS. Aqueous alteration mapping in Rishabdev ultramafic complex using imaging spectroscopy. Int J Appl Earth Obs Geoinf. 2020;88:102084.

Chapter 7

Trans_Proc: A Processor to Implement the Linear Transformations on the Image and Signal Processing and Its Future Scope

Atri Sanyal and Amitabha Sinha

Abstract

We present here Transproc, a reconfigurable generic processor which can execute operations related to linear transformations like FFT, FDCT or FDWT. A graph theoretic lemma is used to find the applicability of such a processor to calculate the flow graph related parallel operations found in these linear transformations. The architecture level design and processing element level design is presented. The primitive instruction set and the control signal implementing the instruction set is proposed. A detailed simulation validating the correctness of PE level and the architecture level data calculation and routing operations are carried out using Xilinx Vivado Webpack. The result related to size, power and timing requirement is presented.

Keywords: Transform processor, Graph Theoretic Concept, Design, Primitive Instruction Set, Simulation

1. Introduction

In this paper we have proposed an efficient architecture for implementation of frequently used and computationally intensive linear transformations in signal or image processing. The linear transformations like FFT, FDCT or FDWT are computationally intensive and also critical for the processing applications. The papers proposing different designs in this domain are mainly of three types. The first category papers propose architectures to implement only a single category of linear transformations like FFT or FDCT [1–14]. Since these implementation's primary focus is on speed so they are mainly implemented on ASIC. These include a variety of algorithms to decrease the number of computationally intensive operations. We have seen multiplier less variety, high speed pipeline, data forwarding, step lifting techniques implementing FFT or FDCT algorithms which greatly decrease the computational complexity and increase the speed, and others. The second category of papers propose processors or architectures which can implement a number of general linear transformations like FFT, FDCT, FDWT. Since these architectures include basic building blocks common to all these transformations and so they need to reconfigure itself before executing different transformations, they are mainly implemented using reconfigurable architectures like FPGA [5–17]. Our paper proposes a processor of that category. The third category of papers discuss implementation of more generic image/signal applications [18–20]. While describing a linear transform data flow graph is used extensively in different literatures. It was proved earlier in [21, 22] by graph theoretical and mathematical induction that a MIMD processor consisting of processing elements connected like a completely connected equi- vertex bi partite graph can copy any actions shown in the flow graph of transformations like FFT, FDCT, FDWT etc. of any arbitrary size. This confirms that a processor with such type of architecture can execute the transforms represented using flow graph method. The architecture of processing element and the overall architecture discussed in [21, 22] is described thoroughly here. The architecture of control unit and the data exchange procedure between the main CPU and memory and this processor and its local memory is discussed in detail here. The instruction set for processing element and the overall processor are all described along with their corresponding control lines. The representative examples of each category of the instruction set are considered and the step wise control signal to implement them is discussed. The entire architecture requires reconfigurability as it is capable of implementing several transforms by its own. Then the architecture is coded in VHDL, synthesized and simulated using Xilinx Vivado. The processor is simulated to verify the operations in three stages. First the component inside the processing element (floating point adder and multiplier) is simulated and tested. Then the longest sequence of execution required in Loefflers FDCT algorithm is tested for each and every processing element and finally the testing of the overall architecture and the data routing between different processing element is simulated and tested. The synthesis result showing the size of the architecture in LUT level and the synthesis result of power and time are discussed. The rest of the paper is composed in this way, Section 2 discusses the theoretical background of the architecture, Section 3 discusses the implementation of the processor in a modular way, the overall architecture of the processor and the implementable CU is presented, then the processing element level architecture is presented, instruction set and the control signals implementing some representative examples of the instruction set is shown. Section 4 discusses the step by step synthesis and the simulation results in terms of speed, timing and size. Finally Section 5 discusses the conclusion and future scope of the work.

2. Proof of the architecture using graph theoretic approach

The theoretical proof using mathematical induction is given in [21] in detail. Here in this paper we will just present a brief of the argument.

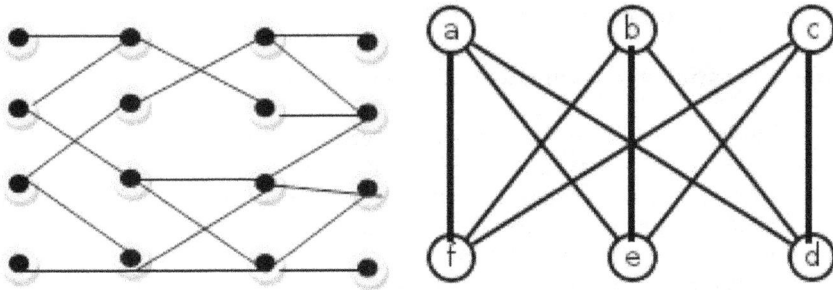

The flow graph shown in the picture [23] is a widely used method of calculating transformations like FFT, FDCT and FDWT. In FFT or FDCT we can see that the flow graph looks physically like an equi vertex k partite graph where k is equal to the no of stages, the vertices are processing elements and the connections among the processing elements are the edges. Since the stages are mutually exclusive among each other so an equi vertex k partite graph like architecture can be reproduced by a fully connected equi vertex bipartite graph if the vertex set contains an one to one mapping between every stage of the k – partite and two stages of the bipartite graph. So any algorithm which is described by a flow graph of the first category can be described by a graph of the later category since the vertex set has the one to one mapping as described. From this argument it is clear that an architecture representing the second category will be efficient as a transform processor and the reconfigurity will make it easy to switch over from one transform to another making it a general transform processor. The orginal architecture requires two sets of processing elements in both the parts and a fully connected bidirectional communication wire between them. The hardware cost can be largely reduced if instead of that we take one set of processing elements and another set of registers, a fully connected feed forward network from register to processing elements and a single feedback network connecting each processing element to their corresponding register. Then the data exchange between two processing elements Pi →Pj can be rewritten as Pi→Ri→Pj. This will take two clock pulses rather than one but the hardware cost will be significantly reduced.

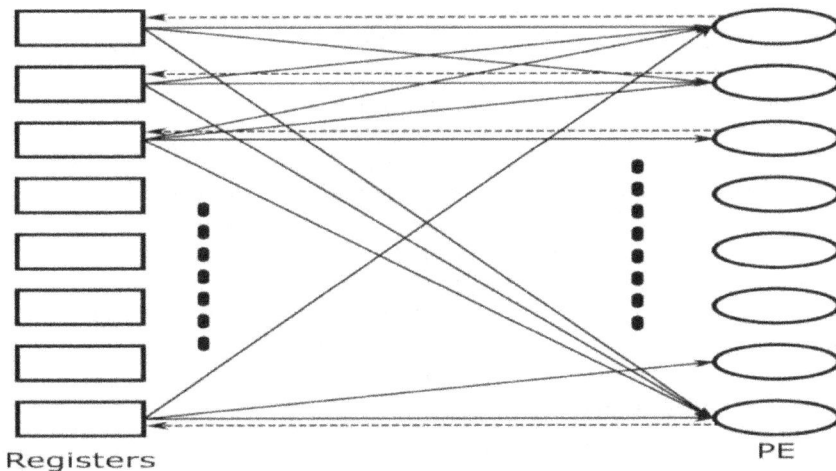

Registers   PE

3. Implementation of the architecture

3.1 Implemenation of the overall architecture design

The fully connected feed forward path described in the previous section is created by 8 multiplexers of size 8 x 3. Each one of them can take input from any eight registers and send the output to any one Processsing Element. The signal lines of the individual multiplexer select the input register loading the value in the Processing Element (PE). This constitutes the most simple but effective feedforward communication lines between the registers and PEs The feedback line is implemented by a combination of 1x2 demultiplexer and 2x1 multiplexer duo which direct the output of the PE to the Input line of the corresponding register.n The same duo can also load the data from the memory in the beginning and once the calculation is complete can store them. The current design is examined with 8 such stages keeping mainly the view of implementing one stage of a FDCT algorithm. The architecture uses 8 bit register sets to latch value while entering or exiting to/from processing element.

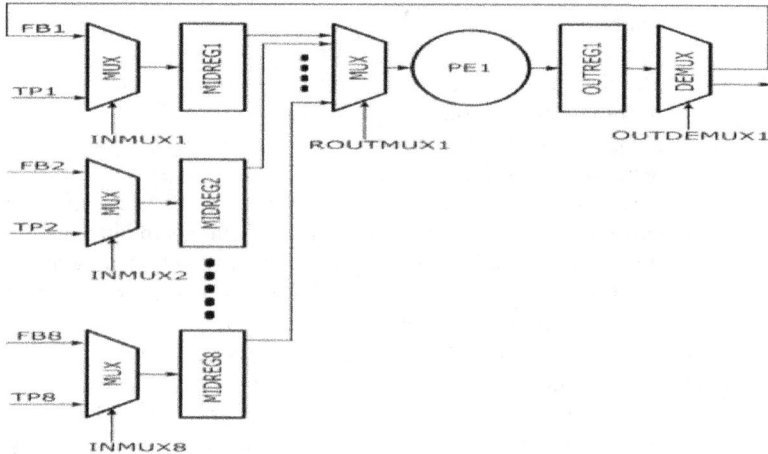

3.2 Implementation of the processing element inside the processor

The implementation of the processing element (PE) inside the processor is done keeping in mind the type of operations which are performed to compute these type of transformations. Most of the operations are floating point type. So we used one floating point adder/subtractor and one floating point multiplier inside the PE. We have used commonly found floating point adder and multiplier in this PE. Keeping open the testing of state of the art designs to improve the performance of adder and multiplier in future for this design. There are two registers which will be used to latch source data of adder and similar two registers which will be used to latch source data of multiplier. The result of adder and multiplier is stored in similar two registers. The PE contains multiplexer and demultiplexer inside to route the data from one internal register to another and to send/receive data to/from the registers outside PE.

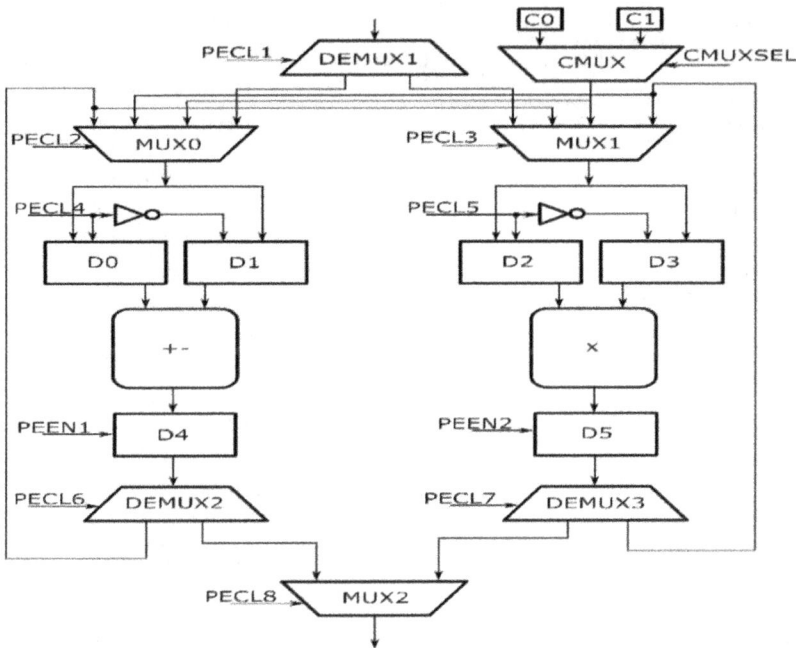

The **Table 1** lists below the routing control signals and their functions for the processor and the routing and activation control signals and their functions for Processing elements PE.

3.3 Primitive instruction set of the processor

The primitive instruction set which is formulated for the processor is mainly contains two categories. Category A is for the instructions to implement routing operations of the processor outside PE and category B is for the arithmetic calculation and data movement operations inside PE.

A. Data Loading/Routing Operations Outside PE:

1. Load Direct MIDREG i: To load data from outside memory in the MIDREG i from TP i [i = 1...8]

2. Load Feedback MIDREG i: To load data in MIDREG i from feedback line FB i [i = 1...8]

3. 3. Rout PE i,MIDREG j = Routing data from any MIDREG j to any PE i. [i,j = 1...8]

4. Out OUTREG i = For storing the value from OUTREG i to outside memory. [i = 1....8]

B. Data Loading/Movement and Mathematical Operations Inside PE:

1. Load [D0-D3][PE i] = to load data in any of the registers D0-D3 from outside memory of PE i.

Signal Name	Select bits	Function
Inmux 1–8	0/1	0 = select input data from outside memory (TP1–8) 1 = select input data from the feedback line (FB1–8)
Routmux1–8	000–111	000 = Select data from the Midreg1 to PE1–8 .... 111 = Select data from the Midreg8 to PE1–8
Outdemux1–8	0/1	0 = select output data from Outreg1–8 to FB1–8 1 = select output data from Outreg1–8 to output
CMUXSEL	000–111	Select any constants C0-C7 based upon the select line.
PECL1	0/1	0 = DEMUX1→MUX0 1 = DEMUX1→MUX1
PECL2	00–11	00 = Direct Load from outside to D0/D1 01 = Movement of data from D4 to D0/D1 10 = Movement of data from D5 to D0/D1 11 = Load constant data from C0-C7
PECL3	00–11	00 = Direct Load from outside to D2/D3 01 = Movement of data from D5 to D2/D3 10 = Movement of data from D4 to D2/D3 11 = Load constant data from C0-C7
PECL4	0/1	0 = Enable bit for D0, 1 = Enable bit for D1
PECL5	0/1	0 = Enable bit for D2, 1 = Enable bit for D3
PEEN1	0/1	1 = Enable bit for D4
PEEN2	0/1	1 = Enable bit for D5
PECL6	0/1	0 = DEMUX2→.MUX0/MUX1 1 = DEMUX2→.MUX2
PECL7	0/1	0 = DEMUX3→.MUX0/MUX1 1 = DEMUX3→.MUX2
PECL8	0/1	0 = Select input from DEMUX2 1 = Select input from DEMUX3

Table 1.
Name of the control signals, there values and functions used in Trans_Proc.

2. Load [D0-D3], [C0-C7], [PE i] = to load data in any of the registers of D0-D3 from any of the constant registers C0-C7of PE i.

3. Add [PE i] = to add the data present in D0 and D1 and keep it in D4 of PE i.

4. Mul [PE i] = to multiply the data present in D2 and D3 and keep it in D5 of PE i.

5. Move [D0-D3], [D4-D5], [PE i] = data movement operation from any of the output registers of D4-D5 to any of the input registers of D0-D3 of PE i.

6. Out [D4-D5], [PE i] = Write back data from any of the output registers D4-D5 of PE i to OUTREG i of PE i.

Next we calculate the total number of instruction per PE and the overall architecture in the **Table 2** below for each group as well as the overall total:

We can see that the total numbers of instructions are 472 out of which 48 are for each PE and 88 are for outside PE. The control signals of the different components

Group name	Total number of instruction per PE	Total number of instruction
A1	N/A	8
A2	N/A	8
A3	N/A	64
A4	N/A	8
B1	4	32
B2	32	256
B3	1	8
B4	1	8
B5	8	64
B6	2	16
Total	48	472

Table 2.
Total no of instructions of different group.

and their functions of the processor units are specified in the previous table, from that we can specify the sequence of control signals which will be activated in order to implement each of the instructions of the instruction set. We can see one representative instruction for each group and the corresponding control signals and their sequence of activation to implement the instruction in the following **Table 3**. The table listing all the instructions can be found in appendix.

3.4 Implementation of operations using the instruction set of the architecture

If we consider the flow graph of the FDCT algorithm of figure taken as an example, we can see that the algorithm is divided into 4 stages and each stage contains 8 PE executing operations which are of three types: floating point addition/subtraction, floating point multiplication and floating point operation evaluating expression of the type $C1*X + C2*Y$. Next, we see a stage wise operation schedule of the 8 PEs (specifying what each PE does in these 4 stages) in the below **Table 4**:

Category A	Sequence of control signals
Load direct MIDREG 1	1.TP1→Data 2. INMUX1→0 3. EN-MIDREG1→1
Load Feedback MIDREG1	1.OUTDEMUX1→0 2. INMUX1→1 3.EN-MIDREG1→1
Rout MIDREG3,PE5	1.EN-MIDREG3→0 2. ROUTMUX5→011
Out OUTREG6	1.EN-OUTREG6→0 2. OUTDEMUX6→1
Category B	Sequence of control signals
Load D0,PE 1	1.input_PE1→data 2.PECL1_PE1→0 3.PECL2_PE1→00 4.PECL4_PE1→1
Load D0,C5,PE 4	1.EN-C5_PE4→1 2.CMUX_PE4→101 3.PECL2_PE4→11 4.PECL4_PE4→1
Add PE3	1.PEEN1_PE3→1
Mul_PE2	2.PEEN2_PE2→1
Out D4_PE7	1.PECL6_PE7→1 2.PECL8_PE7→0 3.data_PE7→output
Move D1,D5,PE 6	1.PECL7_PE6→0 2.PECL2_PE6→10 3. PECL4_PE6→0

Table 3.
Sequence of operations for implementing $C1*X + C2*Y$.

Stage 1	Stage 2
P1: Reg0 + Reg7	P1: Reg0 + Reg3
P2: Reg1 + Reg6	P2: REg1 + Reg2
P3: Reg2 + Reg5	P3: Reg1-Reg2
P4: Reg3 + Reg4	P4: Reg0-Reg3
P5: Reg3-Reg4	P5: $C3\pi/16^*Reg4 + S3\pi/16^*Reg7$
P6: Reg2-Reg5	P6: $C\pi/16^*Reg5 + S\pi/16^*Reg6$
P7: Reg1-Reg6	P7: $-S\pi/16^*Reg5 + C\pi/16^*Reg6$
P8: Reg0-Reg7	P8: $-S3\pi/16^*Reg4 + C3\pi/16^*Reg7$
Stage 3	**Stage 4**
P1: Reg0 + Reg1	P1:——
P2: Reg0-Reg1	P2:——
P3: $\sqrt{2} C3\pi/8^*Reg2 + S3\pi/8^*Reg3$	P3:——
P4: $-S3\pi/8^*Reg2 + \sqrt{2}C3\pi/8^*Reg3$	P4:——
P5: Reg4 + Reg6	P5: Reg4-Reg7
P6: Reg5-Reg7	P6: $\sqrt{2}^*Reg5$
P7: Reg4-Reg6	P7: $\sqrt{2}^*Reg6$
P8: Reg5 + Reg7	P8: Reg4 + Reg7

Table 4.
Stage wise operation schedule 8 PEs performing FDCT algorithm.

We will list the instructions required to execute three cases as a representative example: a > stage 1 operation of PE 5 b > stage 4 operation of PE 7 and c > stage 2 operations of PE 6. These three cases exhibit three category of floating point operations described previously (**Table 5**).

Time unit	Instruction	Description
1	Load direct MIDREG3	Load data from the TP3 line to MIDREG 3.
2	Rout PE[5],MIDREG[3]	Load data from MIDREG3 to input line of PE5
3	Load [D0],[PE 5]	Load input data to D0 from input line of PE 5
4	Load direct MIDREG4	Load data from the TP4 line to MIDREG 4.
5	Rout PE[5],MIDREG[4]	Load data from MIDREG4 to input line of PE5
6	Load [D1],[PE 5]	Load input data to D1 from input line of PE 5
7	Add [PE5]	Add the content of D0 and 2's complement value of D1 and store the value in D4 of PE5
8	Out D4	Output data from D4 to OUTREG 5 of PE 5
9	Load Feedback MIDREG [5]	Load the data from the OUTREG 5 of PE5 to FB5 and then to MIDREG 5.
1	Rout PE[7],MIDREG[6]	Load data from MIDREG6 to input line of PE7
2	Load [D2],[PE 7]	Load input data to D2 from input line of PE 7
3	Load [D3],[C7],[PE 7]	Load D3 with constant from the constant register C7 selected by CMUX
4	Mul [PE7]	Multiply the content of D2 and D3 and store the value in D5 of PE 7
5	Out D5 [PE7]	Output data from D5 to OUTREG 7 of PE 7
6	Load Feedback MIDREG [7]	Load the data from the OUTREG 7 of PE7 to FB7 and then to MIDREG 7.
1	Rout PE[6],MIDERG[5]	Load data from MIDREG5 to input line of PE6
2	Load [D2], [PE 6]	Load input data to D2 from input line of PE 6
3	Load [D3],[C5],[PE 6]	Load D3 with constant from the constant register C5 selected by CMUX
4	Mul [PE 6]	Multiply the content of D2 and D3 and store the value in D5 of PE 6

Time unit	Instruction	Description
5	Mov [D0],[D5],[PE 6]	Move the content from D5 to D0 of PE6
6	Rout PE[6],MIDERG[6]	Load data from MIDREG6 to input line of PE6
7	Load [D2], [PE 6]	Load input data to D2 from input line of PE 6
8	Load [D3],[C6],[PE 6]	Load D3 with constant from the constant register C6 selected by CMUX
9	Mul [PE 6]	Multiply the content of D2 and D3 and store the value in D5 of PE 6
10	Mov [D1],[D5],[PE 6]	Move the content from D5 to D1 of PE6
11	Add [PE 6]	Add the content of D0 and D1 and store the value in D4 of PE6
12	OUT [D4], [PE 6]	Output data from D4 to OUTREG 6 of PE 6
13	Load Feedback Data MIDREG[6]	Load the data from the OUTREG 6 of PE6 to FB6 and then to MIDREG 6.

Table 5.
List of instructions for a > stage 1 operation of PE 5 b > stage 4 operation of PE 7 and c > stage 2 03 operations of PE 6.

3.5 Implementation of the control unit of the processor

Hardwired implementation of the correct control signals, their values and the sequence for total 472 instructions is very difficult physically. Here in this work we have only developed instructions required for proving the correctness of the design, which are of three type. 1. We have developed instructions inside the PE to do a floating point addition and multiplications. 2. We have developed instructions to implement the longest sequence of the FDCT algorithm C1*X + C2*Y inside one PE implemented of a single stage. And 3. Next we have done the same implementation of stage 2 for all PEs and routed the output values randomly to prove the correctness of the implementation. So the control unit is partially developed. We require a programming based approach to develop a full grown assembler to generate all the instructions for all the instructions. These is an incomplete design of the TransProc which we presented in the paper but shows that it has the capability which can be used correctly for generationg all the instructions required for all the transform generators as a hardware co processor implemented in FPGA once the CU is finished generating all the instructions.

4. Simulation and synthesis

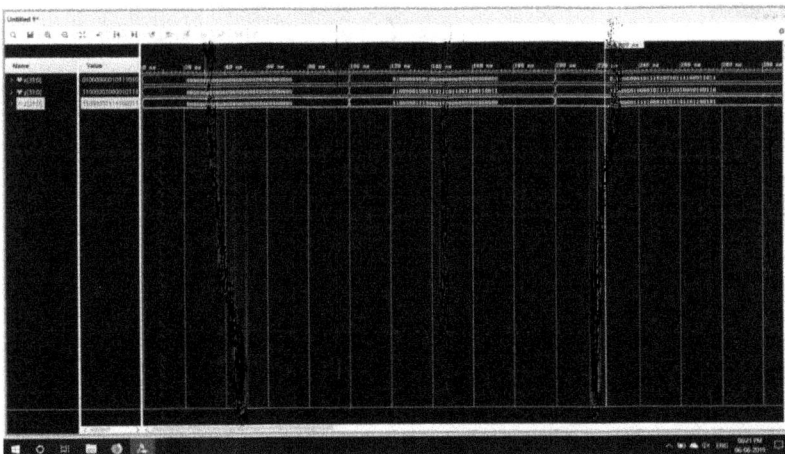

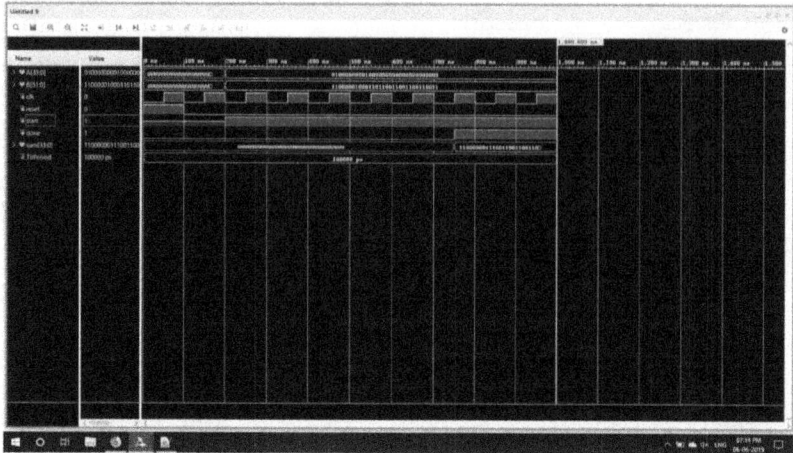

The first two simulations show the correct floating point implementation of floating point multiplier and adder/subtractor. While the floating point multiplier has lots of scope of improvement but floating point adder/subtractor is quite state of the art.

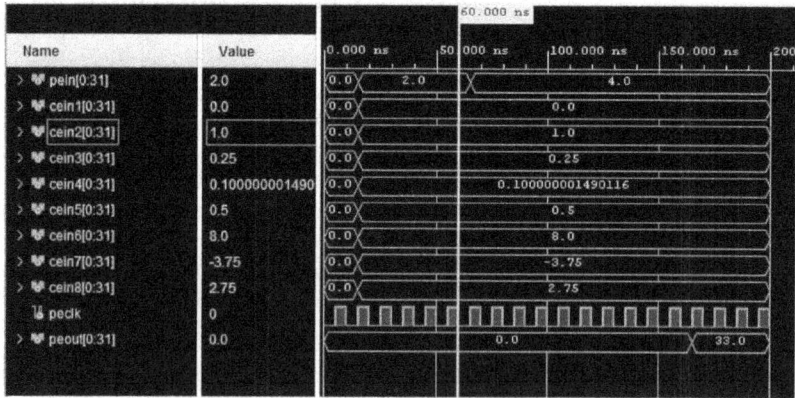

Here we see the longest sequence of multiplication and adder inside a single PE. Pein1xCein5 + Pein1xCEin6 = 2.0x0.5 + 4.0x8.0 = 33.0.

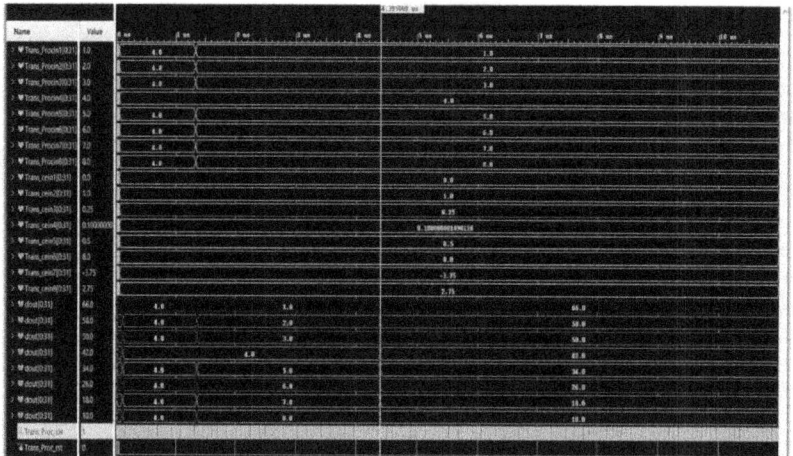

Here we see the routing correctness of the every PEs of the Trans_Proc according to the following flow graph shown in a tabular format:

PE1 = 1, PE2 = 2, PE3 = 3, PE4 = 4, PE5 = 5, PE6 = 6, PE7 = 7, PE8 = 8.
C1 = 2, C2 = 8.
PE1 = PE1x C1 + PE8xC2 = 66.
PE2 = PE2x C1 + PE7xC2 = 58.
PE3 = PE3x C1 + PE6xC2 = 50.
PE4 = PE4x C1 + PE5xC2 = 42.
PE5 = PE4x C2 + PE5xC1 = 34.
PE6 = PE3x C2 + PE6xC1 = 26.
PE7 = PE2x C2 + PE7xC1 = 1.
PE8 = PE1x C2 + PE8xC1 = 10.

This is the way the routing correctness among the different PEs of the processor is tested and we can see that it is working.

Once the behavioral simulation is correctly shown, next we present the result the synthesis of the entire processor done by the Xilinx Vivado and comment on the result (**Tables 6–9**).

The overall utilization report gives an idea of the size of the processor while the number of primitive blocks used in the processor is also given. Please remember that the study here did not include the CU utilization as that is incomplete but and will be used as an separate design in the future study. Total on chip power with its

Utilization report (Summery)	
No of LUT	10897
No of FF	6928
No of IOB	562

Table 6.
Summery of utilization report.

Utilization report (Primitive blocks)		
Primitive name	Number	Functuional category
LUT6	6096	LUT
LUT5	920	LUT
LUT4	2984	LUT
LUT3	576	LUT
LUT2	536	LUT
LUT1	1257	LUT
FDCE	3616	Flop & Latch
FDRE	3312	Flop & Latch
MUXF7	320	MuxFx
CARRY4	168	Carry Logic
IBUF	489	IO
OBUF	73	IO
BUFG	1	Clock

Table 7.
Utilization report of primitives block.

Power report (Summery)	
Total On-Chip Power	0.417 W
Device Dynamic Power	0.335 W
Device Static Power	0.082 W

Table 8.
Power report summery.

Timing report (Summery)	
Max Setup Time	3.419 ns
Worse Pulse Width Slack	4.650 ns
Avg CP required for FP operations inside PE	4
Max Clock Frequency	292 MHz

Table 9.
Timing report summery.

two components dynamic and static is also suggesting an implementable design. T ming report shows Setup up time, WPWS is 4.650 ns, we calculated by hand that the instruction inside the floating point operations inside the takes maximum 4 clock pulses. This makes the maximum clock frequency as 292 MHZ.

5. A discussion on the memory and instruction exchange between the main processor and Trans_proc

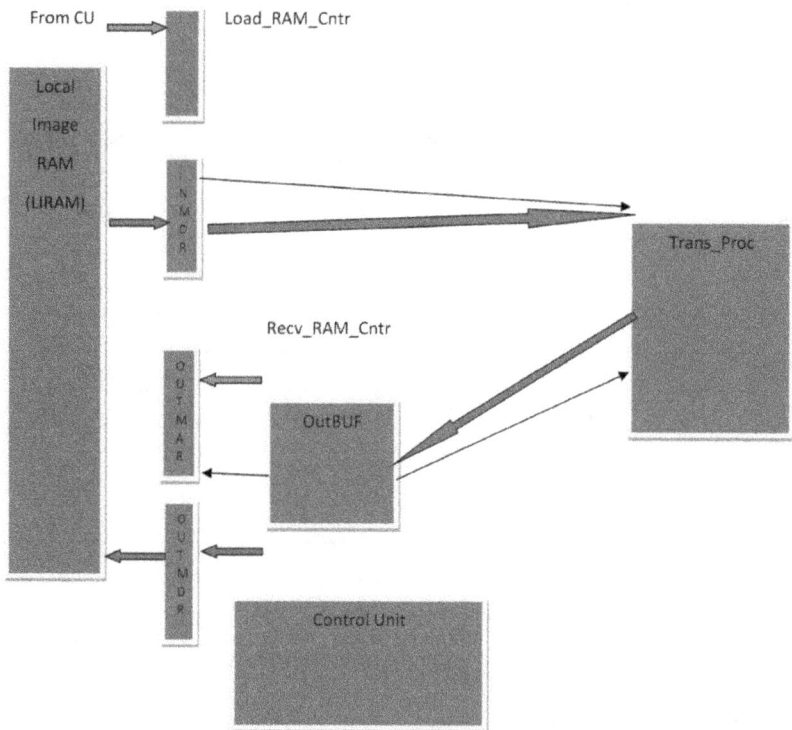

Here we can see the data transfer procedure between the main processor and Trans-Proc which will be implemented as a future scope of this study. The process uses an linear image RAM (LIRAM) to store the primary data. Then there are two data registers used as buffers while going in and out to the Trans-Proc. There is one counter to count the no of blocks going to Trans-Proc and one address register to store the block of transformed image again back to LIRAM. This will be implemented further as the future scope of this study.

Author details

Atri Sanyal[1*] and Amitabha Sinha[2]

1 Amity Institute of Information Technology, Amity University, Kolkata, West Bengal, India

2 UGC Adjunct Faculty, Maulana Abul Kalam Azad University of Technology, West Bengal, India

*Address all correspondence to: atri.sanyal@gmail.com

IntechOpen

References

[1] Po-Chih Tseng et al, "Reconfigurable discrete cosine transform processor for object-based video signal processing", in ISCAS '04. Proceedings of the 2004 International Symposium on Circuits and System, 2004.

[2] Po-Chih Tseng, Chao-Tsung Huang, Liang-Gee Chen, "Reconfigurable Discrete Wavelet Transform Processor for Heterogeneous Reconfigurable Multimedia Systems", Journal of VLSI signal processing systems for signal, image and video technology, 2005.

[3] Gregory W. Donohoe, "The Fast Fourier Transform on a Reconfigurable Processor", Proc. NASA Earth Sciences Technology Conference, Pasadena, CA, June 11-13, 2002

[4] Srivatsava P S V, SaradaV, **"Reconfigurable MDC Architecture Based FFT Processor"**, International Journal of Engineering Research & Technology, 2014

[5] K. Joe Hass David F. Cox, "Transform Processing on a Reconfigurable Data Path Processor", 7th NASA Symposium on VLSI Design 1998

[6] V. Sarada, T. Vigneswaran, "Reconfigurable FFT Processor – A Broader Perspective Survey", International Journal of Engineering and Technology (IJET) 2013

[7] Asadollah Shahbahrami, Mahmood Ahmadi, Stephan Wong, Koen Bertels, "A New Approach to Implement Discrete Wavelet Transform using Collaboration of Reconfigurable Elements", Proc. 2009 International Conference on Reconfigurable Computing and FPGAs

[8] Konstantinos E. Manolopoulos, Konstantinos G. Nakos, Dionysios I. Reisis and Nikolaos G. Vlassopoulos, "Reconfigurable Fast Fourier Transform Architecture for Orthogonal Frequency Division Multiplexing Systems", 2003, available: https://pdfs.semanticscholar. org/dd5c/263725af00e5dd4d42d573c 269f57d917c8d.pdf?_ga=2.84059166.64 0751657.1573804365-914446569. 1569299704

[9] Amitabha Sinha, Mitrava Sarkar, Soumojit Acharyya, Suranjan Chakraborty, "A Novel Reconfigurable Architecture of a DSP Processor for Efficient Mapping of DSP Functions using Field Programmable DSP Arrays", ACM SIGARCH Computer Architecture News Vol. 41, No. 2, May 2013

[10] Sumit Wadekar, Laxman P. Thakare, Dr. A.Y. Deshmukh, "Reconfigurable N-Point FFT Processor Design For OFDM System, International Journal of Engineering Research and General Science Volume 3, Issue 2, March-April, 2015

[11] Alexey Petrovsky, Maxim Rodionov and Alexander Petrovsky, "Dynamic Reconfigurable on the Lifting Steps Wavelet Packet Processor with Frame-Based Psychoacoustic Optimized Time-Frequency Tiling for Real-Time Audio Applications", Design and Architectures for Digital Signal Processing, available: http://www.intechopen.com/books/ design-and-architectures-fordigital-signal-processing2013.

[12] Sharon Thomas & V Sarada, "Design of Reconfigurable FFT Processor With Reduced Area And Power", ITSI Transactions on Electrical and Electronics Engineering (ITSI-TEEE), 2013.

[13] Uma Rajaram, "Design Of Fir Filter For Adaptive Noise Cancellation Using Context Switching Reconfigurable EHW Architecture", Ph.D dissertation, Anna University, Chennai, 2009, available: https://shodhganga.inflibnet.ac.in/ handle/10603/27245

[14] P. S. Reddy, S. Mopuri and A. Acharyya, "A Reconfigurable High Speed Architecture Design for Discrete Hilbert Transform," in *IEEE Signal Processing Letters*, vol. 21, no. 11, pp. 1413-1417, Nov. 2014, doi: 10.1109/LSP.2014.2333745

[15] Atri Sanyal, Swapan Kumar Samaddar, Amitabha Sinha, "A Generalized Architecture for Linear Transform", Proc. IEEE International Conference on CNC 2010, Oct 04-05, 2010, Calicut, Kerala, India, IEEE Computer society, pp. 55-60, ISBN: 97-0-7695-4209-6.

[16] A. Sanyal, S. K. Samaddar, "A Combined Architecture for FDCT Algorithm," Proc. 2012 Third International Conference on Computer and Communication Technology, Allahabad, 2012, pp. 33-37, doi: 10.1109/ICCCT.2012.16

[17] Atri Sanyal, SaloniKumari, Amitabha Sinha, "An Improved Combined Architecture of the Four FDCT Algorithms", International Journal of Research in Electronics and Computer Engineering, (IJRECE), Vol 6 Issue 4 December 2018, ISSN: 2348-2281

[18] Davide Rossi, Fabio Campi, Simone Spolzino, Stefano Pucillo, Roberto Guerrieri, "A Heterogeneous Digital Signal Processor for Dynamically Reconfigurable Computing", IEEE Journal of Solid-State Circuits,Volume: 45, Issue: 8, Aug. 2010

[19] Sohan Purohit, Sai Rahul Chalamalasetti, Martin Margala WimVanderbauwhede, "Throughput/Resource-Efficient Reconfigurable Processor for Multimedia Applications", IEEE Transactions on Very Large Scale Integration (VLSI) Systems, Volume: 21, Issue: 7, July 2013

[20] Vikram, K.N., Vasudevan, V. "Mapping Data-Parallel Tasks Onto Partially Reconfigurable Hybrid Processor Architectures", IEEE Transactions On Very Large Scale Integration (VLSI) Systems, Vol. 14, No. 9, September 2006.

[21] Atri Sanyal, Amitabha Sinha, "A Reconfigurable Architecture to Implement Linear Transforms of Image Processing Applications", International Conference on Frontiers in Computing and System (COMSYS 2020), Jalpaiguri, West Bengal, India, January 13-15,2020

[22] B. Heyne, C. C. Sun, J. Goetze, S. J. Ruan, "A Computationally Efficient High-Quality Cordic Based DCT", 14th European Signal Processing Conference (EUSIPCO 2006), Florence, Italy, September 4-8, 2006

[23] N. Deo, "Graph Theory with applications to engineering and computer science", PHI, 2007

Chapter 8

Application of UAV Remote Sensing in Monitoring Banana Fusarium Wilt

Huichun Ye, Wenjiang Huang, Shanyu Huang, Chaojia Nie, Jiawei Guo and Bei Cui

Abstract

Fusarium wilt poses a current threat to worldwide banana plantation areas. To treat the Fusarium wilt disease and adjust banana planting methods accordingly, it is important to introduce timely monitoring processes. In this chapter, the multispectral images acquired by unmanned aerial vehicle (UAV) was used to establish a method to identify which banana regions were infected or uninfected with Fusarium wilt disease. The vegetation indices (VIs), including the normalised difference vegetation index (NDVI), normalised difference red edge index (NDRE), structural independent pigment index (SIPI), red-edge structural independent pigment index ($SIPI_{RE}$), green chlorophyll index (CI_{green}), red-edge chlorophyll index (CI_{RE}), anthocyanin reflectance index (ARI), and carotenoid index (CARI), were selected for deciding the biophysical and biochemical characteristics of the banana plants. The relationships between the VIs and those plants infected or uninfected with Fusarium wilt were assessed using the binary logistic regression method. The results suggest that UAV-based multispectral imagery with a red-edge band is effective to identify banana Fusarium wilt disease, and that the CIRE had the best performance.

Keywords: Fusarium wilt, banana, UAV, remote sensing

1. Introduction

Bananas (*Musa spp.*) are a widely cultivated cash crop in both the tropical and subtropical regions. Caused by the soilborne fungus *Fusarium oxysporum* f. sp. *cubense* (*Foc*), Banana Fusarium wilt (also known as Panama disease) seriously threatens global banana cultivation and export [1, 2]. As reported, banana Fusarium wilt may have affected up to 100,000 hectares of banana plantations. Moreover, it continues to spread, through infected plant materials and contaminated soil and flowing water, or through farm machinery and inappropriate sanitation measures [2]. Externally, the first sign that a plant is infected with the disease is the withered plant, with the old leaves turning yellow on the edge. With the progression of the disease, the leaves eventually droop and form a 'skirt' around the pseudo-stem before finally falling off. The new leaves may show irregular and wrinkled blades as well as pale margins [3]. Currently, no effective chemical treatment method has been proposed to control Fusarium wilt. "Removal in time" is the optimal way to prevent the disease spread once a diseased plant is identified [4].

For treatment of the disease, and for crop planting adjustments, real-time monitoring and effective identification of banana Fusarium wilt play a significant role [5]. Traditionally, soil investigations have been the only effective means to monitor crop diseases, but such surveys take a lot of time and are often expensive. Recent years have witnessed the rapid development of the remote sensing technology, which has developed into a viable method for disease assessment and monitoring. The leaf pigment content, leaf area index (LAI) and water content of a plant which is infected with a disease will all undergo changes. And such biochemical and biophysical changes in the plant will also present in its spectral reflectance characteristics [6]. Remote sensing technology has been applied to monitor diseases, including Fusarium head blight [7, 8], rust infection [9–11], and powdery mildew [7, 8, 12, 13] in wheat, grey leaf spot in maize [14], bacterial leaf blight in rice [15, 16], and late blight disease and bacterial spot in tomatoes [17, 18] in some studies. However, the sensitivity of spectral bands and VIs varies with the category of diseases. For example, Bravo et al. [19] calculated the normalised difference vegetation index (NDVI) using wavelengths of 620–640 nm and 740–760 nm for extracting powdery mildew from wheat patches. Devadas et al. [20] distinguished yellow rusted wheat from healthy wheat using the anthocyanin reflection index (ARI). Huang et al. [10] suggested that the position of the red edge can be used as a disease indicator. With this in mind, it is of essence to identify which spectral bands and VIs are suitable for the identification of which specific diseases.

UAV remote sensing technology has been developed rapidly over recent years. It has become of interest due to its advantages of long flight time, real-time image transmission, effective detection of high-risk areas, low cost and easy manoeuvrability. It provides new means for the timely and non-destructive extraction of infected plants from the in-season crops [21]. Using UAV multispectral and hyperspectral images, a great number of studies have achieved significant progress in growth monitoring, crop classification, and the identification of diseases and insect pests [22–24]. Within banana production, a few studies have adopted UAV-based images to map the spatial patterns of photosynthetic activity in banana plantations [25]. Nonetheless, there are few studies that use UAV-based remote sensing to monitor banana Fusarium wilt [26, 27]. Furthermore, the spatial scale for remote sensing information and scaling remains one of the fundamental problems in geoscience [28]. Selecting an optimal spatial scale for remote sensing imagery plays a significant role in agricultural monitoring in particular.

Therefore, the goals of this chapter are to: (i) develop an identification method for Fusarium wilt based on UAV multispectral remote sensing, (ii) determine the optimal VI needed for the establishment of a quality identification model, and (iii) evaluate how different image resolutions affect the accuracy of Fusarium wilt identification in order to provide guidance for the application of satellite-based data in a massive scale.

2. Materials and methods

2.1 Field experiment

The experiments were carried out at two experimental locations in Guangxi and Hainan, respectively.

The Guangxi experiment site is located in Guangxi Province of China (23°7′53″ to 23°8′4″ N, 107°43′45 to 107°44′7″ E) (**Figure 1**). It has a subtropical monsoon

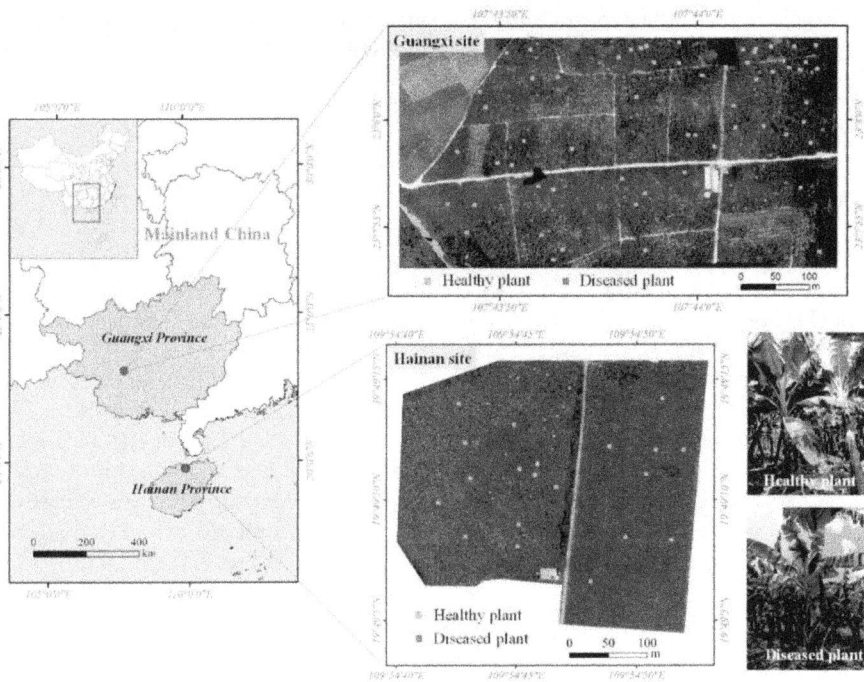

Figure 1.
Location of the experimental sites with the survey sites.

climate characterised by year-round sunshine and rainfall, with a mean annual temperature between 20.8 to 22.4°C, and an average annual rainfall of 1200 mm. The soil type according to the FAO soil classification system is Ferralsol [29]. The banana variety in the study area was "Williams B6". The leaf number of this variety is 34–36, the plant height is about 2.4–3 m, and the growth period is 10–12 months. The banana plantation was established in September 2015, with the planting distance of 2.0 m by 2.6 m. The first harvest was carried out in November 2016. As of August 2018 (the time of the field investigation discussed in this chapter), the third generation of bananas was in the fields and more than 40% of the banana plants were infected with Fusarium wilt.

The Hainan experiment site is located in Hainan Province, China (19°49′4″ to 19°49′16″ N, 109°54′40″ to 109°54′53″ E) (**Figure 1**). It has a tropical monsoon climate characterised by year-round sunshine and rainfall, with a mean annual temperature between 23.1 to 24.5°C and an average annual rainfall of 1750 mm. The soil type according to the FAO soil classification system is Humic Acrisol [29]. This experimental field was divided into two sub-fields (left area and right area) with the middle road as the boundary (**Figure 1**). The left area was developed in June 2017, with the planting distance of 2.0 m by 2.3 m. The first harvest was carried out in July 2018. The banana variety was "Baxijiao". the plant height of this variety is about 2.6–3.2 m and the growth period is 9–12 months. In this field, the rate of banana Fusarium wilt infection was about 10%.

The right area was developed in August 2018. The planting distance was the same as that in the left field. The banana variety was "Nantianhuang". The plant height of this variety is about 2.5–3.0 m and the growth period is 10–13 months. At the time of the field investigation in December 2018, no banana plants were found to be infected with Fusarium wilt.

In this chapter, the experimental data obtained from the Guangxi site was used for calibration and validation of the Fusarium wilt identification model, and from the Hainan site used for model validation.

2.2 Field investigation

2.2.1 Plant investigation

The experiment at Guangxi site was carried out on August 7, 2018. A total of 120 sample plots were investigated to assess the occurrence or non-occurrence of Fusarium wilt (**Figure 1**). Among them, there were 57 healthy samples and 63 diseased samples. The size of each sample plot encompassed one banana plant. Eventually, 75% samples were randomly extracted and employed for the construction of Fusarium wilt identification model denoted by modelling dataset (MD); and the remaining 25% for model validation, denoted by validation dataset 1 (VD1). The experiment at Hainan site was performed on December 11, 2018. The survey strategy was in line with that of the experiment at Guangxi site. A total of 35 sample plots were finally investigated, of which 16 were healthy and 19 were diseased. All the sample plots from Hainan sties were served for model validation, denoted by validation dataset 2 (VD2).

2.2.2 UAV multispectral imagery acquisition

The surveys were carried out by a DJI Phantom 4 Pro quadcopter (DJI Innovations, Shenzhen, China) equipped a MicaSense RedEdge-M multispectral camera (MicaSense, Inc., Seattle, WA, USA). The camera is configured with five bands: Blue (475 nm center, 20 nm bandwidth), Green (560 nm center, 20 nm bandwidth), Red (668 nm center, 10 nm bandwidth), Red edge (717 nm center, 10 nm bandwidth), Near-IR (840 nm center, 40 nm bandwidth). The flight experiment at the Guangxi site was performed between 12:30 p.m.–13:30 p.m. on 7 August 2018, covering an area of 21 ha. While the flight experiment at the Hainan site was implemented between 11:00 a.m.–12:00 p.m. on December11, 2018, covering an area of 11 hectares. The flight altitude above ground level was 120 m with an 8 cm ground sample distance (GSD). Then, the original UAV imagery was resampled to generate images with five resolutions (i.e., 0.5-m, 1-m, 2-m, 5-m, and 10-m) by using nearest neighbour resampling algorithm.

2.3 Data analysis

2.3.1 Vegetation indices

In this section, the VIs method was applied to assess the infection status of Fusarium wilt in banana plantations. Eight VIs that related to plant growth and pigment absorption were selected to characterise the biophysical and biochemical variations due to individual infections. These VIs included the NDVI, normalised difference red edge index (NDRE), structural independent pigment index (SIPI), red-edge structural independent pigment index ($SIPI_{RE}$), green chlorophyll index (CI_{green}), red-edge chlorophyll index (CI_{RE}), anthocyanin reflectance index (ARI), and carotenoid index (CARI). **Table 1** lists the formulations of the VIs.

2.3.2 Statistics analysis

The binary logistic regression (BLR) was used to established the relationships between the VIs and the plants infected or uninfected with Fusarium wilt. As one

VI	Formulation	Sensitive Parameter	Reference
NDVI	$(R_{NIR}-R_{red})/(R_{NIR} + R_{red})$	Green biomass, LAI	[30]
NDRE	$(R_{NIR}-R_{RE})/(R_{NIR} + R_{RE})$	Green biomass, LAI	[31]
SIPI	$(R_{NIR}-R_{blue})/(R_{NIR} - R_{red})$	Leaf pigment content	[32]
SIPI$_{RE}$	$(R_{RE}-R_{blue})/(R_{RE} - R_{red})$	Leaf pigment content	[33]
CI$_{green}$	$R_{NIR}/R_{green}-1$	Leaf chlorophyll content	[34]
CI$_{RE}$	$R_{NIR}/R_{RE}-1$	Leaf chlorophyll content	[35]
ARI	$1/R_{green}-1/R_{RE}$	Leaf anthocyanin content	[36]
CARI	$R_{RE}/R_{green}-1$	Leaf carotenoid content	[37]

Table 1.
List of the VIs used in this chapter.

of the most common multivariate analysis methods, BLR has a dependent variable as a binary variable that represents the presence or absence of an event. The BLR dependent variable is a probability function, which can be expressed as [38]:

$$p = 1/(1+e^{-y}) \qquad (1)$$

where p represents the probability of Fusarium wilt occurrence in this chapter, ranging between 0 and 1, e is the numerical constant, and y refers to the linear combination. They can be expressed in a formula as:

$$y = \beta_0 + \beta_1 x_1 + \beta_2 x_2 + ... + \beta_n x_n \qquad (2)$$

where β_0 refers to the intercept, β_i and x_i (i = 0, 1, 2, ..., n) are the slope coefficients and independent variables, respectively. The logistic regression models were fitted with the modelling dataset through SPSS 20.0 software (SPSS Inc., Chicago, Illinois, USA) in this section.

Following the model fitting, the validation datasets were used to verify the accuracy of Fusarium wilt identification models, with indicators of the Kappa coefficient and overall accuracy (OA) [39, 40]. The Kappa coefficient ranges between −1 and 1, kappa ≥ 0.75 represents excellent agreement, 0.75 > kappa ≥ 0.4 represents fair to good agreement, kappa <0.4 represents poor represents [41]. The OA is the sum of the correctly identified plots divided by the total number of plots.

3. Banana fusarium wilt recognition

3.1 Statistical characteristics of VIs change after disease infection

Table 2 shows the VI values of the diseased and healthy sample plots. Significant differences (independent t-test) were observed in the NDVI, NDRE, CI$_{green}$, CI$_{RE}$, ARI, and CARI values between the healthy plots and diseased plots ($p < 0.01$), but not observed in the SIPI and SIPI$_{RE}$ values ($p > 0.05$). Hence, we selected NDVI, NDRE, CI$_{green}$, CI$_{RE}$, ARI, and CARI for follow-up analysis.

3.2 Recognition model fitting with different VIs

In this section, the relationships between the VIs and the plants infected or uninfected with Fusarium wilt were described by using the BLR method with

Experiment Site	VI	Sample plot	Mean	Std. Deviation	p Value (t-test)
Guangxi site	NDVI	Healthy	0.54	0.11	0.00
		Diseased	0.34	0.14	
	NDRE	Healthy	0.20	0.08	0.00
		Diseased	0.02	0.09	
	SIPI	Healthy	0.88	0.36	0.24
		Diseased	1.68	5.26	
	$SIPI_{RE}$	Healthy	0.58	0.71	0.25
		Diseased	2.07	9.77	
	CI_{green}	Healthy	1.08	0.32	0.00
		Diseased	0.43	0.33	
	CI_{RE}	Healthy	0.56	0.22	0.00
		Diseased	0.09	0.22	
	ARI	Healthy	0.85	0.15	0.00
		Diseased	0.62	0.16	
	CARI	Healthy	0.34	0.04	0.00
		Diseased	0.30	0.06	
Hainan site	NDVI	Healthy	0.44	0.05	0.00
		Diseased	0.36	0.06	
	NDRE	Healthy	0.35	0.10	0.00
		Diseased	0.12	0.09	
	SIPI	Healthy	1.07	0.07	0.06
		Diseased	1.18	0.12	
	$SIPI_{RE}$	Healthy	1.11	0.11	0.04
		Diseased	1.23	0.16	
	CI_{green}	Healthy	0.92	0.26	0.00
		Diseased	0.49	0.26	
	CI_{RE}	Healthy	0.35	0.10	0.00
		Diseased	0.12	0.09	
	ARI	Healthy	0.87	0.30	0.03
		Diseased	0.61	0.35	
	CARI	Healthy	0.43	0.16	0.01
		Diseased	0.33	0.19	

Table 2.
Statistical characteristics of the VI values of the diseased and healthy sample plots.

dataset MD. The classification accuracy of the relational models was verified via both dataset VD1 and VD2. It was found that the use of the NDVI, NDRE, CI_{green}, and CI_{RE} led to relatively good fitting recognition models with the OA values greater than 80% (**Table 3**). Of all the VIs, CI_{RE} obtained the highest verified OA and Kappa coefficient for both VD1 (91.7% for OA and 0.83 for Kappa) and VD2 (80.0% for OA and 0.59 for Kappa), thereby indicating that CI_{RE} performed best in the identification of Fusarium wilt. It could be seen that those VIs containing red-edge

VI	Recognition model	Dataset VD1		Dataset VD2	
		OA (%)	Kappa	OA (%)	Kappa
NDVI	$y = 5.373–11.851 \times NDVI$	83.3	0.66	62.9	0.22
NDRE	$y = 1.802–15.775 \times NDRE$	87.5	0.75	65.7	0.39
CI_{green}	$y = 3.118–4.144 \times CI_{green}$	87.5	0.74	74.3	0.47
CI_{RE}	$y = 1.935–6.110 \times CI_{RE}$	91.7	0.83	80.0	0.59
ARI	$y = 5.326–7.247 \times ARI$	83.3	0.66	68.6	0.37
CARI	$y = 3.172–9.966 \times CARI$	66.7	0.35	60.0	0.21

Table 3.
Recognition models of banana fusarium wilt for different VIs.

band (e.g., NDRE vs. NDVI and CI_{RE} vs. CI_{green}) obtained higher verified OA and Kappa coefficients. Nonetheless, CARI and ARI achieved relatively low verified OA and Kappa coefficients.

3.3 Recognition model fitting with different resolution images

Evaluating the impact of image resolutions on the accuracy of Fusarium wilt recognition can provide guidance for the large-scale application of satellite-based data. In this chapter, the original UAV images were first resampled to five different spatial resolutions (0.5-m, 1-m, 2-m, 5-m, and 10-m), which were then used for Fusarium wilt monitoring. We calculated both the optimal VI without a red-edge band (CI_{green}) and optimal VI with a red-edge band (CI_{RE}) at different resolutions. **Table 4** lists the results of Fusarium wilt recognition model for the CI_{green} and CI_{RE} VIs at different resolutions. As indicated by the verified results, the CI_{RE} at resolution 0.5-m, 1-m, and 2-m were all obtained the acceptable verified OA (over 70%) and Kappa coefficients (over 0.40). When using the dataset VD1, the verified OA at resolution 0.5-m, 1-m, and 2-m were 91.7%, 79.2%, and 75.0%, respectively, and the Kappa coefficients were 0.83, 0.60, and 0.53, respectively. When using dataset VD2, the verified OA at resolution 0.5-m, 1-m, and 2-m were 85.7%, 74.3%, and 71.4%, respectively, and the Kappa coefficients were 0.71, 0.48, and 0.41, respectively. Despite that, the OA and Kappa coefficients at resolution 5-m and 10-m resolution were relatively low, and their values dropped as the resolution decreased. Moreover, at the same resolution, the accuracy of the CI_{green}-based model for Fusarium wilt recognition was lower than that of CI_{RE}-based model. In fact, the only acceptable result for the CI_{green} was at 0.5-m resolution.

3.4 Fusarium wilt Banana distribution mapping at different resolutions

With the aim to further explore the visual effects of image resolutions, the distribution of Fusarium wilt infected and uninfected areas at the Guangxi site were mapped using different resolution images. CI_{RE}-based and CI_{green}-based Fusarium wilt identification models were respectively used to create the Fusarium wilt distribution maps. As can be seen in **Figures 2** and **3**, the maps with 0.08-m, 0.5-m, 1-m and 2-m resolution show quite similar distributions of the occurrence of Fusarium wilt; however, the maps with 5-m and 10-m resolutions exhibited very little detail. **Table 5** lists the area and percentage of the areas infected with Fusarium wilt at different resolutions. For the maps based on CI_{RE} models, the total areas of Fusarium wilt were between 5.69 ha and 6.59 ha, accounting for 38.2% and 44.3% of the

Resolution	Recognition model	Dataset VD1		Dataset VD2	
		OA (%)	Kappa	OA (%)	Kappa
CI_{RE}					
0.5-m	$y = 1.987–5.826 \times CI_{RE}$	91.7	0.83	85.7	0.71
1-m	$y = 1.645–4.896 \times CI_{RE}$	79.2	0.60	74.3	0.48
2-m	$y = 1.475–4.178 \times CI_{RE}$	75.0	0.53	71.4	0.41
5-m	$y = 1.027–2.854 \times CI_{RE}$	70.8	0.42	65.7	0.30
10-m	$y = 0.761–1.817 \times CI_{RE}$	62.5	0.25	62.9	0.24
CI_{green}					
0.5-m	$y = 3.166–3.946 \times CI_{green}$	87.5	0.75	74.3	0.48
1-m	$y = 2.633–3.266 \times CI_{green}$	75.0	0.51	65.7	0.32
2-m	$y = 2.421–2.936 \times CI_{green}$	75.0	0.51	62.9	0.26
5-m	$y = 1.552–1.862 \times CI_{green}$	66.7	0.35	48.6	0.01
10-m	$y = 1.044–1.158 \times CI_{green}$	58.3	0.18	45.7	−0.01

Table 4.
Recognition models of banana fusarium wilt for the CI_{RE} and CI_{green} at different resolutions.

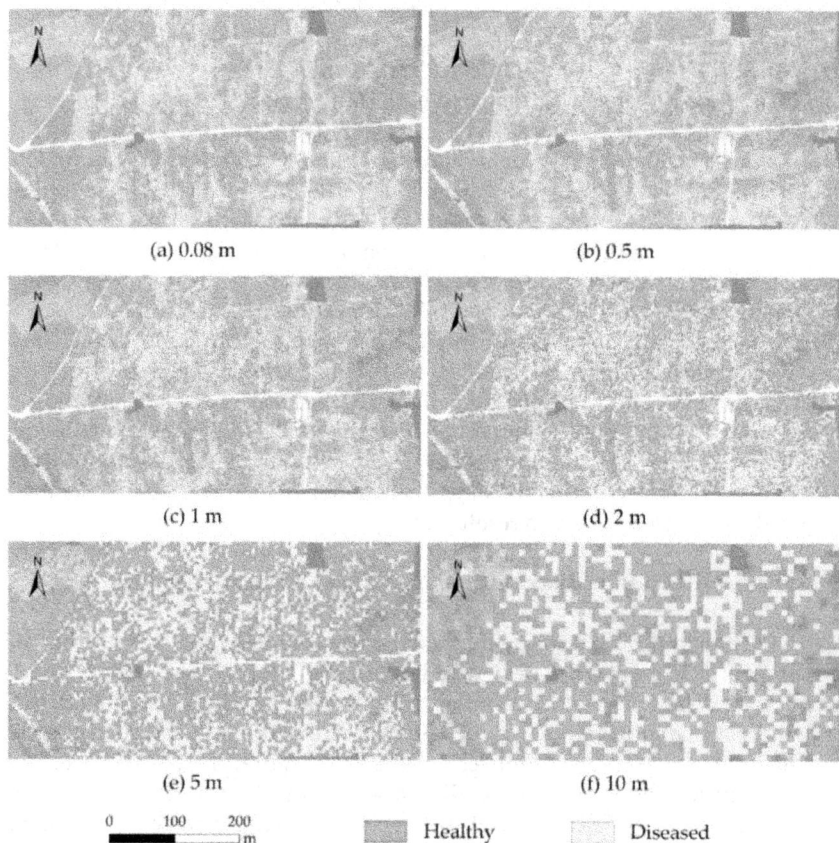

Figure 2.
Maps of the distribution of fusarium wilt based on the CI_{RE} with different resolution images at the Guangxi site.

banana plantation area. Taking a map with a resolution of 2 m as an example, the incidence of Fusarium wilt is between 40.8% and 43.6%. For the maps based on CI_{green} models, the total areas of Fusarium wilt were between 5.09 ha and 6.63 ha, accounting for 34.2% and 44.6% of the banana plantation area. Among them, the percentages of Fusarium wilt of the 0.08-m and 0.5-m resolution maps were 40.1% and 44.6%, respectively.

3.5 Discussion

It was found that among all the VIs used in this chapter, CI_{RE} was the best red-edge VI and CI_{green} was the best non-red-edge VI for Fusarium wilt identification. This is because these two VIs are sensitive to the changes of chlorophyll content of a plant, and Fusarium wilt infection in banana will cause a decrease in leaf chlorophyll content [34, 35, 42]. Furthermore, compared with VIs without the red-edge band, VIs with the red-edge band had higher OA and Kappa coefficients (e.g., NDRE vs. NDVI, and CI_{RE} vs. CI_{green}). It has been widely proved that the red-edge position is very sensitive to the changes of the plant chlorophyll content [43, 44]. Nevertheless, the UAV-based multispectral imagery used in this chapter only possessed 5 bands, which still cannot fully characterise the differences of the spectral characteristics between the diseased and healthy plants. It is therefore of great

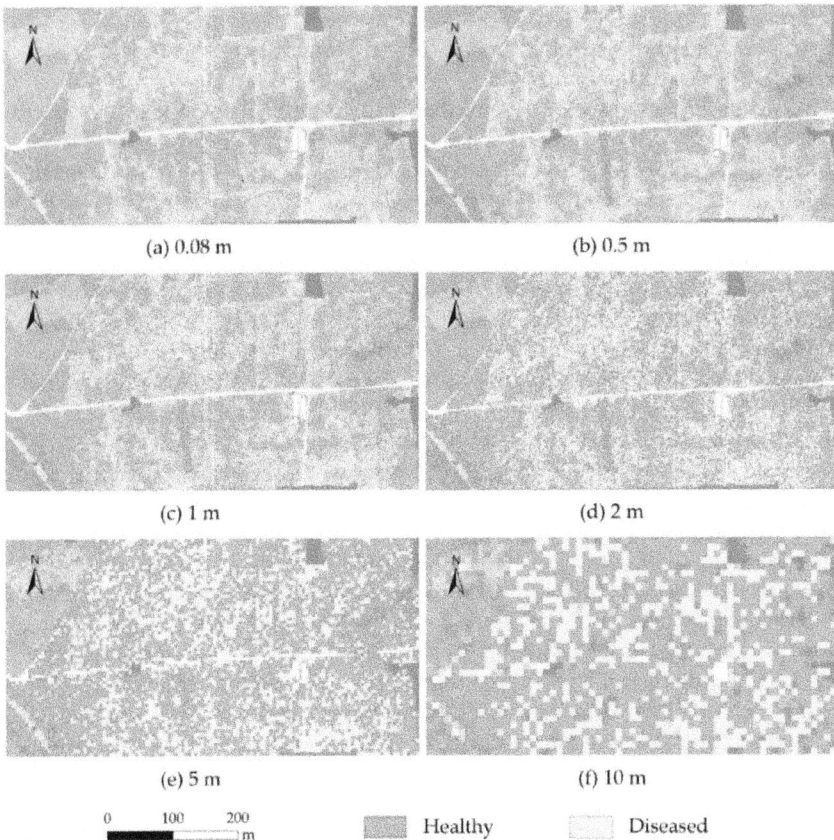

(a) 0.08 m   (b) 0.5 m

(c) 1 m   (d) 2 m

(e) 5 m   (f) 10 m

0   100   200
▬▬▬▬▭▭▭ m   ▨ Healthy   ▨ Diseased

Figure 3.
Maps of the distribution of fusarium wilt based on the CI_{green} with different resolution images at the Guangxi site.

Resolution	Diseased area (ha)	Proportion of diseased area (%)
CI_{RE}		
0.08-m	6.04	40.8
0.5-m	6.59	44.3
1-m	6.28	42.2
2-m	6.47	43.6
5-m	5.70	38.5
10-m	5.69	38.2
CI_{green}		
0.08-m	5.95	40.1
0.5-m	6.63	44.6
1-m	6.44	43.3
2-m	6.63	44.6
5-m	5.69	38.4
10-m	5.09	34.2

Table 5.
Areas of fusarium wilt based on the CI_{RE} and CI_{green} with different resolution images at the Guangxi site.

significance to use hyperspectral data to further study the sensitivity of certain wavebands to banana Fusarium wilt.

The results also showed the potential of combining BLR and VIs to accurately identify Fusarium wilt of banana. Based on this method, an ideal framework for the use of spectral features can be obtained, so as to clarify the pathological mechanisms. In this chapter, the dependent variable was the occurrence of banana Fusarium wilt. Under the circumstance that the predicted variable has a binary nature, BLR can be regarded as a suitable approach [38]. In addition, BLR can deliver better performance than discriminant analysis in the case that the predictor variables are continuous, categorical, or a combination of the two [45]. BLR is highly interpretable, very efficient, and does not require large computational resources, so it is widely used to describe the relationship between a dependent variable and multiple independent variables [38]. Moreover, due to its linear decision surface, non-linear problems cannot be solved by the logistic regression. With the development of artificial intelligence, pattern recognition and machine learning methods will become more common in the use of remote sensing to monitor and predict plant diseases [46].

The Fusarium wilt detection models were verified both using the dataset VD1 VD2. It can be seen from the verification results that both CI_{RE} and CI_{green} performed well in the identification of Fusarium wilt (OA > 70%, and Kappa values > 0.4). This indicates that the detection models of Fusarium wilt have a good transferability in other fields. **Tables 3** and **4** show that the Kappa coefficients of the dataset VD2 were lower than those of the dataset VD1, thus indicating that applying the detection methodology of Fusarium wilt in other fields would cause some precision loss. This situation may be due to the following factors. First of all, one of the most important factors affecting the verification results could be the fact that there were two different banana varieties at the experimental sites ("Williams B6" in VD1 and "Baxijiao" in VD2). These showed that there were differences in their biophysical and biochemical characteristics, which may cause differences in spectral characteristic information. Secondly, due to the differences in the planting time and climatic conditions of the two experimental sites, their growth stages differed greatly. In fact, the banana plants of two experimental areas were at different growth stages during the investigation. Moreover, soil types, planting density, and environmental conditions for crop growth are also important factors that affect the applicability of the Fusarium wilt identification model. Therefore, it is recommended to appropriately optimise the BLR parameters when applying this method in other regions.

In this chapter, the original UAV images were resampled to generate five resolution images (i.e., 0.5-m, 1-m, 2-m, 5-m, and 10-m) to evaluate the impact of different resolutions on the accuracy of Fusarium wilt monitoring. It was found that imagery with a resolution smaller than 2 meters had a good accuracy for Fusarium wilt monitoring, which may be related to the planting spacing and the canopy size of banana. With the reduction of the resolution, the mixed pixel problem influences the precision of object recognition and classification. However, image resolution is not the only difference seen between UAV-based and satellite-based sensors. The wavelength information captured by the satellite-based sensor is different from that of UAV-based sensors. Thus, the simulation results at different resolutions should be further verified with actual satellite-based data. In this chapter, single-period multispectral images were used, which limits the spectral response mechanism to determine the changes in the biophysical and chemical parameters caused by Fusarium wilt. In order to overcome this problem, it is necessary to use multi-temporal and hyperspectral images for dynamic monitoring of the occurrence of Fusarium wilt. Additionally, it is also of great value to explore the differences in the spectral response characteristics of Fusarium wilt and other yellowing stresses (i.e., nutrition deficiency and drought stress).

4. Conclusions

This research used UAV multispectral images to develop a method for identifying Fusarium wilt of banana. The results revealed that the VIs method with BLR analysis can well identify Fusarium wilt. of all the VIs investigated, the CI_{RE} exhibited the optimal performance, with the OA and Kappa coefficients of 91.7% and 0.83 for dataset VD1 and 80.0% and 0.59 for dataset VD2. VIs that included a red edge band obtained better results than those that did not have one. According to the analysis of different resolutions, a resolution smaller than 2 m produced a good identification accuracy of Fusarium wilt. As the resolution decreased however, the identification accuracy decreased. The results indicate that UAV-based multispectral imagery can be applied to identify Fusarium wilt of banana, thus providing reference for disease treatment and crop planting adjustments.

Acknowledgements

This research was funded by Hainan Provincial Major Science and Technology Program of China (ZDKJ2019006); Youth Innovation Promotion Association CAS (2021119); Future Star Talent Program of Aerospace Information Research Institute, Chinese Academy of Sciences (2020KTYWLZX08); National special support program for high-level personnel recruitment (Wenjiang Huang).

Conflict of interest

The authors declare no conflict of interest.

Thanks

We gratefully acknowledge the National Meteorological Information Center of China, Guangxi Jiejiarun Technology Co., Ltd. and Guangxi Jinsui Agriculture Group Co., Ltd. for the experiments.

Author details

Huichun Ye[1,2*], Wenjiang Huang[1,2], Shanyu Huang[3], Chaojia Nie[1], Jiawei Guo[1,4] and Bei Cui[1,2]

1 Key Laboratory of Digital Earth Science, Information Research Institute, Chinese Academy of Sciences, Beijing, China

2 Laboratory of Earth Observation of Hainan Province, Sanya, China

3 Chinese Academy of Agricultural Engineering Planning and Design, Beijing, China

4 School of Marine Technology and Geomatics, Jiangsu Ocean University, Lianyungang, China

*Address all correspondence to: yehc@radi.ac.cn

IntechOpen

References

[1] Shen Z, Xue C, Penton C.R, Thomashow L.S, Zhang N, Wang B, Ruan Y, Li R, Shen Q. Suppression of banana Panama disease induced by soil microbiome reconstruction through an integrated agricultural strategy. Soil. Biol. Biochem. 2019; 128: 164-174. http://dx.doi.org/10.1016/j.soilbio. 2018.10.016.

[2] Ordonez N, Seidl M F, Waalwijk C, Drenth A, Kilian A, Thomma B P H J, Ploetz R C, Kema G H J. Worse comes to worst: Bananas and Panama disease-when plant and pathogen clones meet. PLoS Pathog. 2015; 11: e1005197. http:// dx.doi.org/10.1371/journal.ppat. 1005197.

[3] Van den Berg N, Berger D K, Hein I, Birch P R, Wingfield M J, Viljoen A. Tolerance in banana to Fusarium wilt is associated with early up-regulation of cell wall-strengthening genes in the roots. Mol. Plant Pathol. 2007; 8: 333-341. http://dx.doi.org/10.1111/j. 1364-3703.2007.00389.x.

[4] Lin B, Shen H. Fusarium oxysporum f. sp. cubense. In: Wan F, Jiang M, Zhan A, editors. Biological Invasions and Its Management in China: Volume 2. Springer Singapore: Singapore, Singapore, 2017.

[5] Shi Y, Huang W, Ye H, Ruan C, Xing N, Geng Y, Dong Y, Peng D. Partial least square discriminant analysis based on normalized two-stage vegetation indices for mapping damage from rice diseases using PlanetScope datasets. Sensors 2018; 18: 1901. http://dx.doi. org/10.3390/s18061901.

[6] Zheng Q, Huang W, Cui X, Shi Y, Liu L. New spectral index for detecting wheat yellow rust using Sentinel-2 multispectral imagery. Sensors 2018; 18: 868. http://dx.doi.org/10.3390/ s18030868.

[7] Jin X, Jie L, Wang S, Qi H J, Li S W. Classifying wheat hyperspectral pixels of healthy heads and Fusarium head blight disease using a deep neural network in the wild field. Remote Sens. 2018; 10: 395. http://dx.doi.org/10.3390/ rs10030395.

[8] Mahlein A K, Alisaac E, Al Masri A, Behmann J, Dehne H W, Oerke E C. Comparison and Combination of Thermal Fluorescence and Hyperspectral Imaging for Monitoring Fusarium Head Blight of Wheat on Spikelet Scale. Sensors 2019; 19: 2281. http://dx.doi.org/10.3390/s19102281.

[9] Huang W, Lamb D.W, Niu Z, Zhang Y, Liu L, Wang J. Identification of yellow rust in wheat using in-situ spectral reflectance measurements and airborne hyperspectral imaging. Precis. Agric. 2007; 8: 187-197. http://dx.doi. org/10.1007/s11119-007-9038-9.

[10] Huang W, Guan Q, Luo J, Zhang J, Zhao J, Liang D, Huang L, Zhang D. New optimized spectral indices for identifying and monitoring winter wheat diseases. IEEE J. Sel. Top. Appl. Earth Observ. Remote Sens. 2014; 7: 2516-2524. http://dx.doi.org/10.1109/ JSTARS.2013.2294961.

[11] Shi Y, Huang W, Gonzalez-Moreno P, Luke B, Dong Y, Zheng Q, Ma H, Liu L. Wavelet-based rust spectral feature set (WRSFs): A novel spectral feature set based on continuous wavelet transformation for tracking progressive host-pathogen interaction of yellow rust on wheat. Remote Sens. 2018; 10: 525. http://dx.doi.org/10.3390/ rs10040525.

[12] Yuan L, Pu R, Zhang J, Wang J, Yang H. Using high spatial resolution satellite imagery for mapping powdery mildew at a regional scale. Precis. Agric. 2016; 17: 332-348. http://dx.doi. org/10.1007/s11119-015-9421-x.

[13] Zhao J, Xu C, Xu J, Huang L, Zhang D, Liang D. Forecasting the wheat powdery mildew (Blumeria graminis f. Sp tritici) using a remote sensing-based decision-tree classification at a provincial scale. Australas Plant Path. 2018; 47: 53-61. http://dx.doi.org/10.1007/s13313-017-0527-7.

[14] Dhau I, Adam E, Mutanga O, Ayisi K, Abdel-Rahman E.M, Odindi J, Masocha M. Testing the capability of spectral resolution of the new multispectral sensors on detecting the severity of grey leaf spot disease in maize crop. Geocarto Int. 2018; 33: 1223-1236. http://dx.doi.org/10.1080/10 106049.2017.1343391.

[15] Huang J, Liao H, Zhu Y, Sun J, Sun Q, Liu X. Hyperspectral detection of rice damaged by rice leaf folder (Cnaphalocrocis medinalis). Comput. Electron. Agric. 2012; 82: 100-107. http://dx.doi.org/10.1016/j.compag.2012.01.002.

[16] Yang C M. Assessment of the severity of bacterial leaf blight in rice using canopy hyperspectral reflectance. Precis. Agric. 2010; 11: 61-81. http://dx.doi.org/10.1007/s11119-009-9122-4.

[17] Zhang M, Qin Z, Liu X, Ustin S.L. Detection of stress in tomatoes induced by late blight disease in California USA using hyperspectral remote sensin. Int. J. Appl. Earth. Obs. Geoinf. 2003; 4: 295-310. http://dx.doi.org/10.1016/S0303-2434(03)00008-4.

[18] Jones C.D, Jones J.B, Lee W.S. Diagnosis of bacterial spot of tomato using spectral signatures. Comput. Electron. Agric. 2010; 74: 329-335. http://dx.doi.org/10.1016/j.compag.2010.09.008.

[19] Bravo C, Moshou D, West J, McCartney A, Ramon H. Early disease detection in wheat fields using spectral reflectance. Biosyst. Eng. 2003; 84: 137-145. http://dx.doi.org/10.1016/S1537-5110(02)00269-6.

[20] Devadas R, Lamb D.W, Simpfendorfer S, Backhouse D. Evaluating ten spectral vegetation indices for identifying rust infection in individual wheat leaves. Precis. Agric. 2009; 10: 459-470. http://dx.doi.org/10.1007/s11119-008-9100-2.

[21] Deng L, Mao Z, Li X, Hu Z, Yan Y. UAV-based multispectral remote sensing for precision agriculture: A comparison between different cameras. ISPRS J. Photogramm. Remote Sens. 2018; 146: 124-136. http://dx.doi.org/10.1016/j.isprsjprs.2018.09.008.

[22] Dash J.P, Watt M.S, Pearse G.D, Heaphy M, Dungey H.S. Assessing very high resolution UAV imagery for monitoring forest health during a simulated disease outbreak. ISPRS J. Photogramm. Remote Sens. 2017; 131: 1-14. http://dx.doi.org/10.1016/j.isprsjprs.2017.07.007.

[23] Liu B, Shi Y, Duan Y, Wu W. UAV-based crops classification with joint features from orthoimage and DSM data. Int. Arch. Photogram. Remote Sens. Spat. Inform. Sci. 2018; XLII-3: 1023-1028. http://dx.doi.org/10.5194/isprs-archives-XLII-3-1023-2018.

[24] Liu K, Zhou Q B, Wu W B, Xia T, Tang H J. Estimating the crop leaf area index using hyperspectral remote sensing. J. Integr. Agr. 2016; 15: 475-491. http://dx.doi.org/10.1016/S2095-3119(15)61073-5.

[25] Machovina B L, Feeley K J, Machovina B J. UAV remote sensing of spatial variation in banana production. Crop Pasture Sci. 2016; 67: 1281-1287. http://dx.doi.org/10.1071/CP16135.

[26] Ye H, Cui B, Huang S, Dong Y, Huang W, Ren Y, Guo A, Jin Y. Identification of banana Fusarium wilt disease based on UAV multi-spectral

imagery. In: Proceedings of the International Conference on Intelligent Agriculture, Beijing, China, 18-21 October 2019.

[27] Ye H, Huang W, Huang S, Cui B, Dong Y, Guo A, Ren Y, Jin Y. Recognition of Banana Fusarium Wilt Based on UAV Remote Sensing. Remote Sens. 2020; 12: 938. http://dx.doi.org/10.3390/rs12060938.

[28] Li N, Xie G, Zhou D, Zhang C, Jiao C. Remote sensing classification of marsh wetland with different resolution images. J. Resour. Ecol. 2016; 7: 107-114. http://dx.doi.org/10.5814/j.issn.1674-764x.2016.02.005.

[29] IUSS Working Group WRB. World Reference Base for Soil Resources 2006; FAO: Rome Italy 2006.

[30] Rouse J W, Haas R H, Schell J A, Deering D W. Monitoring vegetation systems in the great plains with ERTS. In: Proceedings of the Third ERTS-1 Symposium NASA SP-351, Greenbelt, MD, USA, 10-14 December 1973.

[31] Gitelson A, Merzlyak M N. Spectral reflectance changes associated with autumn senescence of aesculus–hippocastanum L and acer-platanoides L leaves—spectral features and relation to chlorophyll estimation. J. Plant Physiol. 1994; 143: 286-292. http://dx.doi.org/10.1016/S0176-1617(11)81633-0.

[32] Ozdemir A. Using a binary logistic regression method and GIS for evaluating and mapping the groundwater spring potential in the Sultan Mountains (Aksehir Turkey). J. Hydrol. 2011; 405: 123-136.

[33] Ramoelo A, Skidmore A K, Cho M A, Schlerf M, Mathieu R, Heitkonig I M A. Regional estimation of savanna grass nitrogen using the red-edge band of the spaceborne RapidEye sensor. Int. J. Appl. Earth. Obs. Geoinf. 2012; 19: 151-162. http://dx.doi.org/10.1016/j.jag.2012.05.009.

[34] Gitelson A A, Gritz Y, Merzlyak M N. Relationships between leaf chlorophyll content and spectral reflectance and algorithms for non-destructive chlorophyll assessment in higher plant leaves. J. Plant Physiol. 2003; 160: 271-282. http://dx.doi.org/10.1078/0176-1617-00887.

[35] Gitelson A A, Vina A, Ciganda V, Rundquist D C, Arkebauer T J. Remote estimation of canopy chlorophyll content in crops. Geophys. Res. Lett. 2005; 32: L08403. http://dx.doi.org/10.1029/2005GL022688.

[36] Zhou X, Huang W, Zhang J, Kong W, Casa R, Huang Y. A novel combined spectral index for estimating the ratio of carotenoid to chlorophyll content to monitor crop physiological and phenological status. Int. J. Appl. Earth. Obs. Geoinf. 2019; 76: 128-142. http://dx.doi.org/10.1016/j.jag.2018.10.012.

[37] Zhou X, Huang W, Zhang J, Kong W, Casa R, Huang Y. A novel combined spectral index for estimating the ratio of carotenoid to chlorophyll content to monitor crop physiological and phenological status. Int. J. Appl. Earth. Obs. Geoinf. 2019; 76: 128-142. http://dx.doi.org/10.1016/j.jag.2018.10.012.

[38] Lee S, Pradhan B. Landslide hazard mapping at Selangor Malaysia using frequency ratio and logistic regression models. Landslides 2007 4: 33-41. http://dx.doi.org/10.1007/s10346-006-0047-y.

[39] Congalton R G. A review of assessing the accuracy of classifications of remotely sensed data. Remote Sens. Environ. 1991; 37: 35-46. http://dx.doi.org/10.1016/0034-4257(91)90048-B.

[40] Foody G M. Classification accuracy comparison: Hypothesis tests and the

use of confidence intervals in
evaluations of difference equivalence
and non-inferiority. Remote Sens.
Environ. 2009; 113: 1658-1663. http://
dx.doi.org/10.1016/j.rse.2009.03.014.

[41] Tuanmu M N, Viña A, Bearer S,
Xu W, Ouyang Z, Zhang H, Liu J.
Mapping understory vegetation using
phenological characteristics derived
from remotely sensed data. Remote
Sens. Environ. 2010; 114: 1833-1844.
http://dx.doi.org/10.1016/j.rse.2010.
03.008.

[42] Dong X, Wang M, Ling N, Shen Q,
Guo S. Potential role of photosynthesis-
related factors in banana metabolism
and defense against Fusarium
oxysporum f. sp cubense. Environ. Exp.
Bot. 2016; 129: 4-12. http://dx.doi.
org/10.1016/j.envexpbot.2016.01.005.

[43] Clevers J G P W, de Jong S M, Epema
G F, van der Meer F, Bakker W H,
Skidmore A.K, Addink E.A. MERIS and
the red-edge position. Int. J. Appl.
Earth. Obs. Geoinf. 2001; 3: 313-320.
http://dx.doi.org/10.1016/S0303-2434
(01)85038-8.

[44] Dash J, Curran P J. The MERIS
terrestrial chlorophyll index. Int. J.
Remote Sens. 2004; 25: 5403-5413.
http://dx.doi.org/10.1080/0143116042
000274015.

[45] Mathew J, Jha V K, Rawat G S.
Application of binary logistic regression
analysis and its validation for landslide
susceptibility mapping in part of
Garhwal Himalaya India. Int. J. Remote
Sens. 2007; 28: 2257-2275. http://dx.doi.
org/10.1080/01431160600928583.

[46] Lu J, Sun L, Huang W. Research
progress in monitoring and forecasting
of crop pests and diseases by remote
sensing. Remote Sens. Technol. Appl.
2019; 34: 21-32. http://dx.doi.
org/10.11873/j.issn.1004-0323.
2019.1.0021.

Chapter 9

Satellite Control System: Part II - Control Modes, Power, Interface, and Testing

Yuri V. Kim

Abstract

This part II of the chapter Satellite Control System (SCS) was originally planned for publishing in the book Satellite Systems (Acad. Ed. Dr. T. Nguyen), dedicated to the Systems Design, Modeling, Simulation, and Analysis, together with the Part I (SCS Architecture and Main Components). However, restricted volume of this book did not let the publisher to put then this part in the book. The book Recent Applications in Remote Sensing (Acad. Ed. Prof. M. Marhgany) considers the various aspects of the optical and radiolocation sensing and imaging of the Earth surface from Space. Consequently, as it was presented in the Part I, the author adheres to the point of view here that satellite is not just a platform to carry in Space a payload, but is equipment integration system and its designer is in charge for fully integrated and Space-qualified Space segment, which with the corresponding operation and ground equipment would be capable to successfully execute dedicated mission (Remote Sensing). The material, presented in this part, briefly highlights the basic aspects of SCS control modes, electric and informational interface, and ground testing, which would promote successful interaction with satellite payload, such as Remote Sensing subsystem and mission success.

Keywords: satellite control, attitude and orbit, determination, estimation, sensors, actuators, coordinate systems, reference frame, state estimation and Kalman filtering, earth gravity and magnetic fields, interface, assembling integration and testing (AIT), space qualification

1. Introduction

For a remote sensing satellite, equipped with a remote sensing payload, the Satellite Control system (SCS) is very important, providing for the payload required attitude a position in space.

Hence, the payload functionality and its performance essentially depend on SCS characteristics.

Often, specifically, the payload provider is responsible for satellite system integration and mission success. That is why payload company engineers should be familiar with SCS and its role in the satellite mission performance. Below in the introduction some physical principles, showing dependence of the remote sensing payload characteristics on SCS, are briefly discussed.

Historically radiolocation sensing of underlying Earth surface (footprints) started from the high-altitude air patrol aircrafts, performing the military reconnaissance purposes. They were equipped with special on-board Radio Location Stations (RLS), having Side-Looking Antenna (SAR). The resolution (image quality) of such an RLS is essentially dependent on the SAR available length, which for an aircraft cannot exceed a few meters. With the development of space exploration, special remote sensing satellites became available for the Earth observation, and the military reconnaissance purposes were essentially extended for the civil applications such as exploration of Earth-borne disasters, rescuing, agriculture, forestry, and others. Using for Earth observation space platforms brought to this process new important benefits. The main observations from them are as follows: broader instantaneously observed from the space areas (spot 20–30 km), high resolution (3–5 m), and the capability to observe in relatively short periods of time big areas of Earth with periodically repeatable underlying tracks. Technologically, by developing RLS equipment, new benefits were achieved with using much longer antennas (about 10–15 m) and synthesized analytically aperture (SSAR—side looking synthetic aperture radar). More information about RLS SSAR technology can be found in Refs. [1–6]. Considering Earth remote sensing system, we have to emphasize that successful operation of such a system is not available without physical (stabilization) or analytical availability of satellite angular attitude and its position between SSAR coherent radio pulses, used in system memory for building SSAR analytical aperture.

Any Space country using remote sensing satellites has to take care first about satellite attitude and orbital determination and control or, in other words, about SCS and its precision [7–9].

In **Figure 1** below three generations of Canadian Earth observation satellites are presented, Radarsat-1, Radarsat-2, and Radarsat Constellation (RCM). Canada is a pioneer of using SAR technology for the civil tasks for Earth Remote sensing and since launch of Radarsat-1 in November of 1995 has accumulated a big experience and gained tremendous achievements in this area.

Figure 2 presents the basic SAR principles to get radiolocation image reflected from Earth radio signal.

The satellite with on-board RLS-SAR orbits the Earth with the orbital velocity $\overline{V}$ in the flight direction. The microwave beam is transmitted obliquely at right angles to the direction of flight illuminating a swath. Slant range ρ (reflected signal corresponding time delay) is measured, assuming that the flight altitude h and the look angle and satellite position in orbit are known and nominal. Let us consider SAR with a real aperture antenna, **Figure 3**.

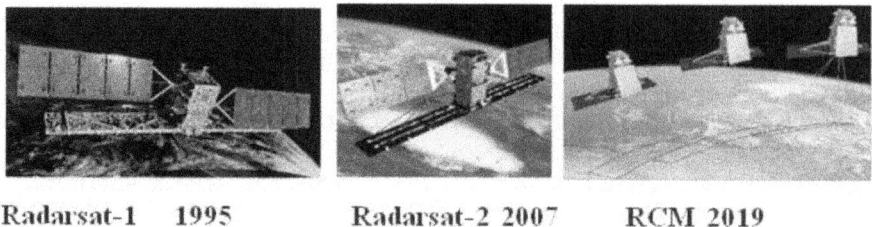

Radarsat-1   1995   Radarsat-2 2007   RCM 2019

Figure 1.
Canadian family of earth observation SAR satellites. Copyright: CSA//www.asc-csa.gc.ca.

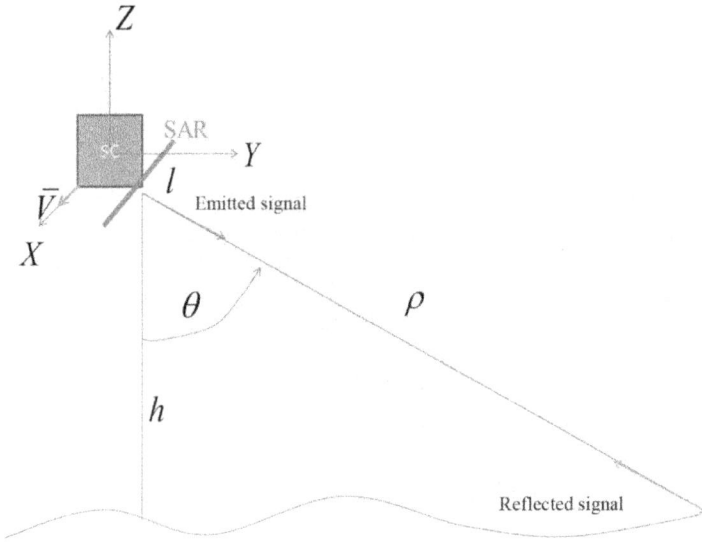

Figure 2.
SAR principles. ρ is slant range, θ is look angle, h is flight altitude, SC is satellite, SAR is RLS SAR antenna, XYZ is satellite body frame, $\bar{V}$ is flight velocity, l is the antenna length.

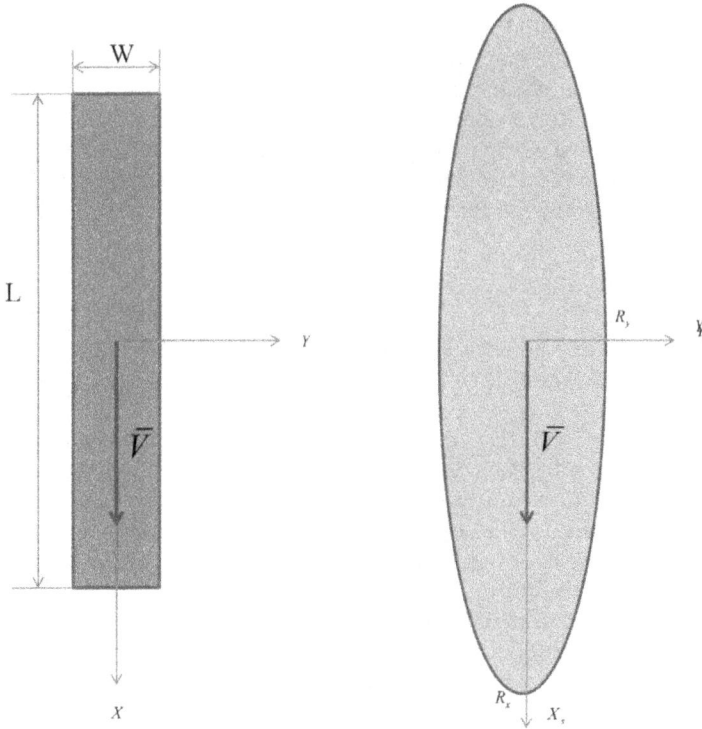

Figure 3.
Left (red): SAR antenna, right (green): SAR beam footprint. L is antenna length, W is antenna width, R_x and R_y are SAR beam footprint dimensions, $\bar{V}$ satellite ground track vector.

Then, SAR beam footprint dimensions in lateral (Y) and longitudinal (X) directions (R_y and R_x resolutions) can be found with the following formulas [1, 4, 6]:

$$R_y = \frac{c\tau}{2\sin\theta_y} = \frac{\Delta\lambda}{2\sin\theta_y} \tag{1}$$

$$R_x = \frac{h\lambda}{L\cos\theta_x} \tag{2}$$

where c is light speed in vaccum ($\sim$300,000 km/s), τ is RLS transmitted pulse duration, λ is RLS transmitted carrier frequency wavelength, $\Delta\lambda$ is RLS transmitted pulse bandwidth, h is flight altitude, θ_x and θ_y are SAR look angles in X and Y directions correspondingly.

For example, for following numerical data (Radarsat-1):

$$L = 15 \text{ m}, W = 1.5 \text{ m}, h = 800 \text{ km}, \theta_x = 85^0, \theta_y = 23^0,$$
$$\lambda = 0.06 \text{ m (C-band)}, \Delta\lambda = 10 \text{ m}$$

$$R_y = \frac{\Delta\lambda}{2\sin\theta_y} = \frac{10}{2\sin 23^0} = 12.8 \text{ m}$$

$$R_x = \frac{h\lambda}{L\cos\theta_x} = \frac{800000 \cdot 0.06}{15\cos 5^0} = 3212 \text{m}$$

As one can see from the numerical example above, *the physical aperture* SAR antenna (15 m) can provide quite a good resolution in the lateral direction, but not good enough in the longitudinal. However, this resolution can be drastically improved with the *synthetic aperture* SAR antenna (SSAR) [1, 4, 6] when all reflected pulses, collected during certain period of time of Earth radiation, are summarized in SSAR RLS on-board computer analytically with the purpose to get RLS image, like it could be created by a physical antenna with a big length.

In this case, much higher longitudinal resolution can be achieved that theoretically can be expressed by the formula:

$$R_x = \frac{L}{2} \tag{3}$$

This formula for the numerical example below provides the longitudinal resolution $R_x = 7.5$ mthat drastically improves the resolution of the same RLS, but with physical SAR aperture.

It should be mentioned that formulas ((1)–(3)) assume a certain steady nominal satellite angular orientation (e.g., zero), set by the three Euler angles: Pitch ($\alpha_y = 0$), Roll ($\alpha_x = 0$), and Yaw ($\alpha_z = 0$) that in turn assume absolutely accurate satellite attitude determination and control. Also, the process of synthesis of the artificial SAR aperture–SSAR assumes absolute accurate knowledge of satellite position. In practice, the attitude and the position are measured and controlled with certain errors that lead to change of SAR RLS resolution and as a result, to the distortion of RLS picture (deterioration of image quality). Not only big steady errors impact on the image quality, but a small jitter also. Therefore, for the SCS of Remote Sensing satellites where the payload is SSAR RLS essential are accuracy requirements that can be transformed in resulted SSAR resolution distortion errors. Analysis of this effect can be found in special literature [3, 10]. Here, we just introduce the reader to some specific SCS tasks that can help for understanding of the integration process of SSAR RLS and SCS on satellite. The reader himself can carry out the impression about the importance of proper installation (mechanical interface) and mutual alignment of the mechanical axes of SAR, satellite bus, GPS antenna, and attitude determination devices (e.g., the Star Tracker). For modern satellites, we can provide some approximate numbers, related to a precise SCS performance (attitude knowledge: $15' - 2''$, attitude control: $0.5° - 1°$, position knowledge: 10–30 m, position control 100–150 m).

Required SCS functionality and the performance must be ensured by a certain design order, prescribed by the System Engineering discipline and validated during Assembling Integration and Test complain in the Space Qualification Functional test. Some of these aspects, namely: Control Modes, Power, Interface, and Testing are covered by this Part II material presented in the book.

2. Typical SCS control modes

SCS dynamics can usually be presented by closed negative feedback control loop, which consists of three typical components: plant (satellite), observer-estimator (sensors and navigation-attitude/orbit determination algorithms), and controller (actuators and control algorithms[1]). Modern approach to its design is analytical synthesis, based on optimal/suboptimal algorithms, provided by System Estimation and Control Theory [11, 12] and Mathematical Model-Based Design tools from the MathWorks Inc. [13].

However, in practice (after evaluation of potentially available optimal solution), conventional engineering design, based on former experience, still has been widely used. This approach brings some generic system (SCS) architecture, components, and operational modes.

Typical SCS (mainly, attitude control ACS) modes are as follows: Idle, Acquisition, Pointing, Maneuvering, and Safe Hold Mode. Mainly, all of them can be activated/deactivated by the ground commands from the Mission Control Center (MCC) and/or automatically (by on-board software). Orbital maneuvering and sometimes attitude (slew) are executed exceptionally by the ground commands.

Anyway, a special command flag is generated upon reception of the TLM mode transition command from MCC or after analyzing some internal SCS flags, generated following the operational logic, time, and system components' state and status.

2.1 Idle

After satellite separation from the launch vehicle before starting AODCS operational modes, it could be in the so-called IDLE mode. It checks the system and its components' state, satellite attitude and angular velocity, and orbit. The system is powered (ON) and activated (operational), except of the actuators. It provides from the sensors TLM data to MCC for operational analysis. Satellite actuators are not controlled, and it has random attitude and free rotation initially initiated by the separation pulse from the launch rocket separation mechanism. If ground analysis confirms the system state is nominal and angular motion is safe to have sufficient electric power and thermal conditions, then SCS actuators can be activated to start satellite control.

2.2 Acquisition

This mode can consist of two phases: de-tumbling and initial acquisition mode.

If premature de-spinning is required and applied, then it is usually performed with MAG and MTR, using B-dot algorithm. Three-axis magnetometer (MAG)

---

[1] Single satellite orbital control is usually executed by the telemetry (TLM) control command from ground.

output provides measured Earth magnetic field induction vector **B**. Its components are differentiated and become proportional to $\dot{B}_{x,y,z}$. With appropriate control gains, these signals are applied to the magnetic torque rods (MTRs). Initial spinning is decelerated and the dumping process is finalized to the slow satellite rotation (a few degrees per sec) about the local vector of Earth magnetic induction **B**. Next starts the acquisition phase with coarse three-axis attitude determination and control. It starts with three-axis attitude determination and PID control (e.g., MAG, Sun Sensor-SS, MTR, and the Reaction Wheels unit—RW [9]). In this mode, satellite body axes *XYZ* are prematurely aligned in parallel with desired reference axes $X_r Y_r Z_r$. This mode is usually fast. Control loop bandwidth in this mode is wide, transfer process termination time is as short as possible, final attitude accuracy is coarse (about of a few degrees). In some applications, this mode can start directly without previous application of the de-spinning sub-mode.

2.3 Pointing

This usually is the operational working mode, required for the successful payload operation. In this mode, the sensitivity axis of satellite payload instrument is accurately pointed in the required direction. Accurate alignment (about a few angular minutes) with the reference frame axes (where the required direction for the payload instrument is set) should be achieved in this mode. The most accurate attitude sensors (e.g., Star Tracer-ST) are applied. The control loop bandwidth can be narrowed to filter external disturbances and measured noise more effectively. In this mode, the control is slow but precise.

2.4 Maneuvering

2.4.1 Orbital

Orbital control thrusters are activated by the computed autonomously on-board or dent from ground TLM command to perform scheduled orbit correction/maneuver. Pre-calculated thruster activation time Δt is used to execute satellite orbit correction pulse $\Delta V = T\Delta t$ (where T is thrusters' force). For example, in the orbital flight direction to increase degraded with time satellite orbit altitude.

2.4.2 Attitude (slew)

In this mode, satellite is controlled in the closed control loop to turn it by the desired angle to point it in new desired direction. Control law in this mode can be as follows:

$$T_c = -k_p(\alpha - \alpha_c) - k_d \dot{\alpha} \tag{4}$$

where T_c is control torque, k_p and k_d are proportional and damping control gains, α and $\dot{\alpha}$ are angular deviation, and the velocity α_c is desired attitude angle.

2.5 Safe hold mode (SHM)

This mode is commanded in some dangerous situations, when satellite life critical failure or flight anomaly is automatically detected on-board by on-board computer software (OBC SW) or identified on ground by satellite operators after TLM data analysis. In this mode, SCS system task is to keep an appropriate satellite angular orientation with respect to the Sun and the Earth providing sufficient thermal, electrical (solar panel energy generation), and radio communication (antennas orientation) conditions for as long as possible time (idealistically *the indefinite* SHM) and consuming as less power as possible (battery electric energy and attitude control thrusters cold gas). Idealistically, a satellite should have *a passive* SHM, when SCS system can be in the state OFF. During the SHM, the Operation Team should resolve the problem that caused this mode transition and start the recovery procedure (transition in the Acquisition mode). Commands to transition in and recovery from SHM can be considered as the final results (command flags) of the special satellite Failure Detection Isolation and Recovery (FDIR) [14] algorithm that can be realized outside of SCS.

3. Electric power and informational interface

3.1 Electric power

Satellite bus using solar panels (SP), on-board rechargeable batteries, and centralized Power Management Unit (PMU) [15] supplies SCS with available DC voltage. For example, 28 V/50 V and power 500 W from one SP at Sun incidence angle <5 deg. If satellite has 2 SP nominally permanently facing sun, then available-electric power is about 1 kW. During Sun eclipse periods and when sunlight is not sufficient for the SP to generate enough electric power, on-board batteries are used. For example, let us assume that two lithium-ion batteries (voltage 28 V, capacity 12 Ah (350 Wh), depth of discharge—DOD = 50% each) are used to provide storage power during Sun eclipse periods, contingency shadowing, or SP failure.

AEU power convertors convert PMU voltage in lower voltages required for SCS OBC, sensors, and actuators (e.g., 3.3 V, 5 V, 12 V). SCS power consumption depends on its current operational mode and may vary from the nominal Pn to the minimum (Pm) value. For example, SCS power budget is as follows:

- OBCS (related to SCS part)—10 W/3 W, GPS-7 W, 3-axis MAG-0.5 W, SS-0 W (photo sensor) ST-8 W, 3xRS-1.2 W, 3xRW(s)-240 W, 3xMTR (s)-60 W.

- Then for the nominal operation Pn = 397.5 W and for the active SHM (only OBCS –(3 W), SS, MAG, MTRs are "ON") Pm = 63.5 W.

- This example shows that satellite can nominally operate with fully lightened SP consuming for SCS approximately 40% of available generated on-board by SP electric power.

- SHM: if only satellite life essential equipment is powered on in this mode and if by some reason Sun direction is totally lost or has not been captured, then AODCS can work for about 6 hours without recharging the batteries.

3.2 Electric interface

Mainly two types of interface are used for informational connection SCS equipment and date exchange, as follows:

1. Analogue interface with separate pair of wires in satellite harness. This type is usually applied within AODCS for simple analog devices thermistors, Sun sensors, horizon sensors, etc.

2. Digital bus lines [15] that are applicable for all satellite digital equipment. This type is used with digital devices that have embedded digital computer such as GPS, ST, etc. Electrical interface for all AODCS equipment is usually is defined in Interface Control Document(s) (ICD), that is, essentially, data exchange protocol defining also signal electrical characteristics, connectors, and pins. It is worth to mention here, at least, two following busses:

   a. MIL-STD-1553B [16] is US military standard that defines a TDM multiple-source-multiple-sink data bus. By definition, MIL-STD-1553B is a *bidirectional*, half-duplex (when transmit cannot receive) deterministic communications protocol with central control (i.e., on-board computer or OBC), where each member (i.e., remote terminal) can receive or transmit data. A 1553B network consists of four major components: transmission media, remote terminals, a bus controller, and a bus monitor. The transmission media is a twisted, shielded wire pair with direct or transformer coupling. The data rate is 1 Mbps of Manchester-encoded, bi-phase data stream. Up to 32 words can comprise a single message in which each word is 20 bits long. One system can accommodate up to 31 remote terminals, a bus controller, and a bus monitor (**Figure 4**).

   b. RS-422 (TIA/EIA-422) [17], as it was named by the American Standard National Institute ANSI, is a technical standard that specifies electrical characteristics of a digital signal circuit used by International Electronic Industry. It is digital, serial, asynchronous, one direction, differential, point-to-point line (1 transmitter and 10 receivers, 10 Mbps) interface.

Figure 5 shows that a differential signaling interface circuit consists of a driver with differential outputs and a receiver with differential inputs.

With using voltage between the wires A and B and the ground, the transmitter transmits and receives serial flow of digital data in the binary form 0/1.

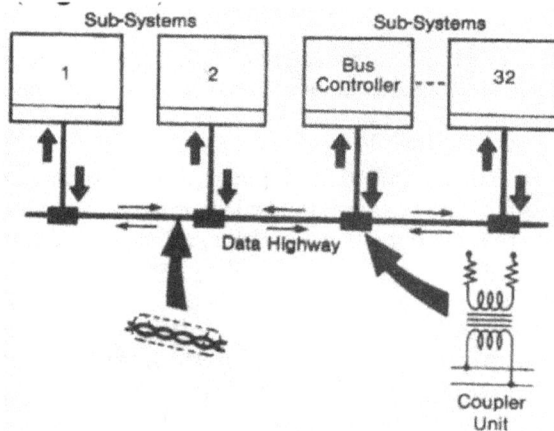

Figure 4.
MIL STD 1553 bus.

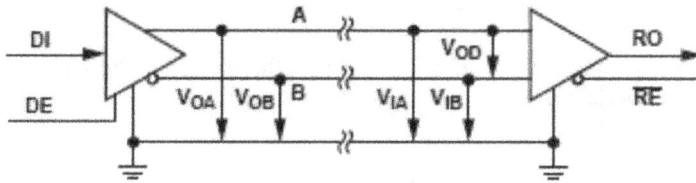

Figure 5.
RS-422 serial bus, lines A and B. Transmitter (driver D), receiver (R), I—input, O—output, V—voltage.

4. SCS space environment protection and testing

The problems with Ground Tests of Space Systems (SS[2]) at first appeared together with the lunch of the first human-made Earth orbiting satellites: Sputnick 1 (1957, USSR, launcher R-7), Explorer-1 (1958, USA, launcher RS-29/Juno), Alouette-1 (1962, Canada, launcher DM-21/Thor-Agena B, USA). Unlike air flight vehicles, flying mainly below an attitude of 25 km in the Earth's athmosphere, space vehicles-SS (under the acronym SS in futher considereation we will undermind Space vechicle (spacecraft-S/C) and their equipment and components—S/C subsystems) should fly in Space at altitude above 225 km, practicaly without atmospheric pressure, in other words in a vacuum, and in addition being affected by the cosmic radiation.

For the air vehicles (airplanes), environmental conditions at that time were already studied and well known, and ground test procedures existed and were almost conventional. But for the SS they were totally new, as well as the launch mechanical impact. Therefore, they were to be studied and ground test types, methodology, and the procedures developed.

By now, it has already been done and presented in many International and National standards and regulations.

After studying space environment and accumulation of some experience with launch and operation of SS, a new group of special ground tests was developed and introduced in the form of related standards and following procedures and documents presented in [8, 18, 19]. This group of tests generally includes the following types: Thermo and Vacuum (TVAC), Vibration and Strength, Radio Communication and Electro Magnetic Compatibility (EMC), final refinement and verification of system assembling and integration (AIT). These tests are finalized by the customer or authorized independent expert conclusion about launch readiness and named Space Qualification (SQ).

4.1 Environmental conditions

SCS system should be designed to work in Space environment conditions that briefly can be characterized by the data below. It has to have special protection measures to satisfy space requirements [8]. It also should be taken into account that different system components may be located inside or outside (SS, ST, HS, Thrusters, GPS antenna) of the satellite and be installed close to the nominally hottest or the coldest surface of its body. However, typically all system components are subjects of the environmental tests [8, 14] to verify different requirements for the internal and external system devices.

---

[2] Under the acronym SS in futher considereation we will undermind Space vechicle (spacecraft-S/C) and their equipment and components—S/C subsystems.

4.1.1 External pressure is close to vacuum [h = 500 km, $P = 10^{-7}$ Pa (1 Pa≈10^{-5}atm)]

Impact: outgassing, change of material strength. Protection: hermetic sealing, application of special materials, filling by the inert gas.

4.1.2 Temperature (no direct contact with Earth atmosphere, hence no heat convection as in Aviation)

The heat balance between S/C and space takes place exclusively because of the particles radiation. The main sources of external radiation are Sun radiation and Earth-infrared reflection (Earth albedo).

Impact: S/C temperature significantly depends on its orientation relatively to Sun and Earth. For example, for a small satellite m = 200 kg, cube: 1 m x 1 m x 1 m), Temperature: Sun side +90C, Space side -20C, Earth side -10C. There are extreme temperature gradients between S/C sides.

Protection: Special thermal design (thermos radiators and plates, materials, and painting) and Thermal Control System (TCS) (thermistors and heaters, convectional ventilation) are applied.

4.1.3 Electromagnetic Space radiation (Special effects in South Atlantic region)

Impact: on electronic equipment (mainly OBC: Glitches, Single Event Upsets, Latch effects). Protection: Application of special radiation-resistant electronic elements, radiation case, special radiation protective shielding.

4.1.4 Disturbing influence of Earth magnetic field, residual atmosphere, and solar pressure

Impact: disturbing forces and torques, affecting satellite orbit and the attitude.

Protection: periodic orbital correction, demagnetization on ground, and minimization of the ballistic coefficient, effective attitude control.

4.1.5 Hard electromagnetic compatibility (EMC) conditions because of small volume for the accommodation

Impact: mutual electromagnetic interference.
Protection: appropriate allocation, screening.

4.1.6 Sun eclipse

Impact: A-Solar panels cannot be used, and satellite power is provided only by the on-board batteries that cannot be discharged during the eclipse period less the critical voltage. AODCS should minimize power consumption.

B-SS cannot be used for the attitude determination.

Protection: A-Application of the sufficient type of batteries, starting eclipse with previously charged prime and redundant batteries.

D-Switching during the eclipse period attitude determination method to another sensor that does not need Sun visibility (e.g., HS) and/or using the Momentum wheel for gyro stabilization of the satellite.

4.1.7 Gravity acceleration

The gravity field acceleration is decreasing with increasing of satellite attitude. It influences on the required velocity for the space flight in circular orbit at this altitude. This velocity can be calculated with the following formula [7]:

$$V_0 = \sqrt{gR} \qquad (5)$$

where g is the gravity acceleration, $R = R_e + h$ is the distance between satellite and the Earth center, $R_e = 6378.137$ km is Earth spherical model radius (equatorial), h is satellite altitude.

Gravity gradient torque T_g impacting on a cylindrical shape satellite attitude is as follows [7]:

$$T_g = -\frac{3}{2}\omega_0^2 \left(J_e - J_p\right)\sin\alpha \qquad (6)$$

where $\omega_0 = \frac{V_0}{R}$ is satellite orbital rate, J_e is satellite equatorial moment of inertia, J_p is satellite polar moment of inertia, α is angle of deviation of satellite from local horizontal plane.

Earth gravity acceleration can be calculated with the following formula [7].

$$g = \frac{\mu}{(R_e + h)^2} \qquad (7)$$

where $\mu = \gamma M_e = 398600.5 km^3/s^2$ is Earth gravity constant, γ is Universal Gravity Constant, and M_e is the Earth mass. Calculated with this formula, gravity acceleration at the Earth surface is $g(R_e) = 9.798 \ m/s^2$. The graph of the gravity acceleration calculated with (7) is presented in **Figure 6**.

4.2 Environmental tests

To verify that SCS does meet the environmental requirements, AODCS usually examines with special environmental tests.

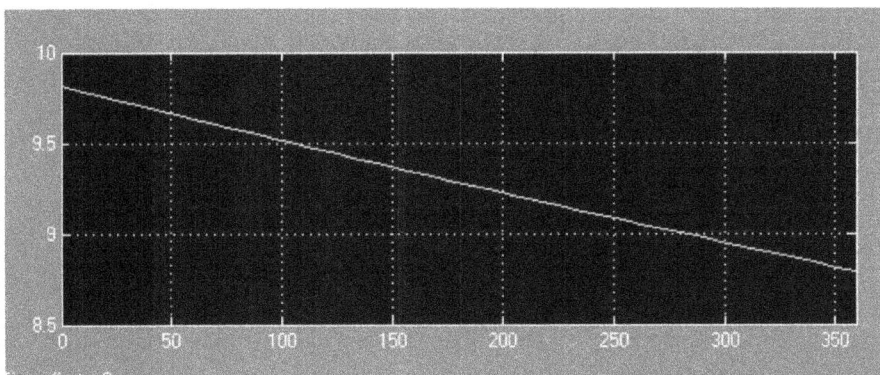

Figure 6.
Gravity acceleration versa altitude g(h) (m/s²), h (km).

The following tests related to environmental conditions could be performed[3]:

1. Thermo, Vacuum (TVAC-cyclic), and Humidity Category.

   a. For internal components: Temperature 20C +/-5C for the unit with thermostat, but -60C – +40—worst case of the thermostat failure.

   b. For external component, for example, -100C–+150C.

2. Pressure: h = 0 km - P = 1 atm (101 kPa); h = 20 km - P = 0.05 atm (5.06 kPa); h = 609.6 km- P $= 4.74 \cdot 10^{-12}$ atm $(4.8 \cdot 10^{-10}$ kPa);

3. Electromagnetic compatibility and interface (EMC/EMI and magnetic cleanness) (depends on system accommodation and RF antenna patterns).

4. Radiation Hardness, Radiation hardiness designators M, D, FG, P, L, H indicate unit capability to withstand to a certain radiation dose, for example, $M = 3 \cdot 10^3$ rad (Si);

5. Mechanical launch impacts: sinusoidal/random/acoustic vibration in a certain range of frequencies, static, and shock (depends on planned launcher).

4.3 Space qualification (SQ) functional test (FT)

Usually, SQ FT is carried out in specially equipped for these purpose facilities by trained personal and highly qualified experts as the final part of system Assembling, Integration, and Testing Activities (AIT).

It must be mentioned that AIT activities should include this final functional test for SS Flight Model that should demonstrate its capabilities to perform in Space required functions after all other type of SQ (environmental) tests under the system have been performed.

In this SQ FT SS is completely assembled and integrated, as well as refined (calibrated). Specifically, in this test SS hardware (H/W) and software (S/W) working jointly should be verified. This test should finalize SQ procedures, preceding release of the Space Qualification Report (SQR), and declaring readiness SS for launch and operation in Space.

Unfortunately, in common practice due to many various reasons, SQ FT does not occupy the right place in a number of SQ tests. For example, for such important for any spacecraft system as Attitude Control System (ACS), this test often comes down to checking electric interface and right direction of rotation of the Reaction wheels ("polarity test"). Sometimes, such a superficial attitude to SQ FT leads to very stressful and even dramatic situations after the launch during SS operation in space. That is why many authors [20–22] draw attention SS developers to this problem and present some simulation tools and procedures to resolve it.

With regard to the satellite control system (SCS) and its components [9, 23], the main difficulty for FSQ is to model on ground orbital flight with relevant gravitation and magnetic field, and orbital motion. For these purposes, for modern

---

[3] Approximate range of changing conditions shown for example.

small satellite, very sophisticated test beds, based on three degrees of freedom air bearing tables, have being used [22].

Here, the author presents a different approach from System Dynamics Identification point of View [24]. This general approach allows to identify SS (in particular, SCS) dynamics in open control loop using currently commonly available for engineers Matlab/Simulink Identification Toolbox. It does not require complex test (control and verification) equipment. Mainly special laboratory emulators, activating SS sensors must be used additionally to conventional AIT SQ equipment (assembling stand, laboratory registration console for simulation radio link to satellite Tracing, Telemetry, and Control System (TTCS), power supply, installation devices, and mass property determination machine).

Looking at the problem of SS SQ FT from the point of view of System Dynamics theory, we can allege that if system has proper dynamics, previously verified with mathematical simulation (MSim), which meets design requirements, and it (structure and parameters) is validated with semi-natural simulation (SNSim); then, this system will be capable to perform expected functions in space, at least in some mission essential operation modes. The process of evaluation of system dynamics by the experimental way is named System Identification process [25]. Presently, identification methods have been developed to be practically used in many engineer applications. The most known and commonly used engineer tool for the identification is Matlab/Simulink Identification Toolbox (ITB) [26]. It is applicable for both cases; when system structure (mathematical model) is partly known and only unknowns are system parameters (mathematical model coefficients)—"gray box" case and when considered system is totally unknown —"black box" case. For both these cases, ITB allows to identify (estimate) system mathematical model. Only experimentally measured system input and output signals are used. The ITB adjusts the most suitable model estimate to minimize the difference between the output measured experimentally and its estimation, provided by the estimated model. Briefly, the essential elements of this identification are presented below. Let us consider an SS as a unit consisting of the harware (HW) and the software (SW) components as presented in **Figure 7**.

From the System Dynamics point of view, this system can be characterized by its input $x(t)$, output $y(t)$ and some mathematical operation, determinning system conversion from the output to the input

$$y(t) = [y(t)] \qquad (8)$$

At the first approximation, many Aerospace devices and systems dynamic can be considered in the scope of Linear Time Invariant (LTI) dynamic system theory. In this case, (8) can be represented as follows:

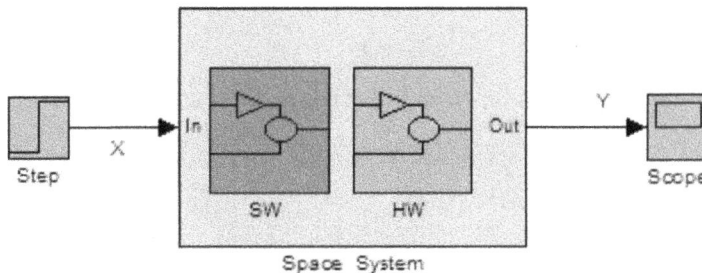

Figure 7.
Space system unit.

$$y(t) = \int_0^t g(t - \tau)x(\tau)d\tau \qquad (9)$$

where $g(t)$ is system's impulse characteristic response to the Dirac's input impulse $x(t) = \delta(\tau)$. Using Laplace transformation to (9), it can be represented as

$$y(s) = G(s)x(s) \qquad (10)$$

where $y(s) = L[y(t)]$, $x(s) = L[x(t)] = L[y(t)]$ are Laplace transformation of output and input signals and $G(s) = L[g(t)]$ is Laplace transformation of system impulse function. In other words,

$$G(s) = \frac{y(s)}{x(s)} \qquad (11)$$

is the ratio of Laplace transformations of output to input signals.

In general case, LTI system transfer function can be expressed as the two polynomial ratios:

$$G(s) = \frac{b_m s^m + b_{m-1} s^{m-1} + b_{m-2} s^{m-2} + \dots \dots + b_1 s + b_0}{a_m s^n + a_{n-1} s^{n-1} + a_{n-2} s^{n-2} + \dots \dots + a_1 s + a_0} \qquad (12)$$

where b_i, a_j are constant polynomial coefficients, $m \leq n$. Usually, (12) represents a stable system with the characteristic equation

$$a_m s^n + a_{n-1} s^{n-1} + a_{n-2} s^{n-2} + \dots \dots + a_1 s + a_0 = 0 \qquad (13)$$

which roots $s_{k1,2} = \mathrm{Re}_k \pm j\mathrm{Im}_k$ satisfy the following condition

$$\mathrm{Re}_k \leq 0 \qquad (14)$$

Usually, for any designed SS assumable (before identification) transfer function $G(s)$ for system unit, presented in **Figure 4**, is known from its design documentation.

Identification experiment provides measured input $X_m(t)$ and output $Y_m(t)$ data (**Figure 8**) and the identification procedures used in ITB allows to estimate this function coefficients $\hat{a}_i$ and $\hat{b}_i$.

Theoretical ratio between the input x and the output y of a LTI system $G(s)$ is (11). However practically, it takes place experimentally measuring input x_m and output data, distorted by some input V_i and output V_o errors

$$x_m = x + V_i \qquad (15)$$

and

$$y_m = y + V_o \qquad (16)$$

The difference between expected and experimental output signals is as follows:

$$e = y_m - y = \hat{G}(s)x_m + V_o - G(s)x = \hat{G}(s)(x + V_i) + V_o - G(s)x =$$
$$= \left[\hat{G}(s) - G(s)\right]x + \hat{G}(s)V_i + V_o \qquad (17)$$

where $\hat{G}(s)$ is estimate of system transfer function.

This difference (14) is used in ITB to tune (adjust) model coefficients $\hat{a}_i$ and $\hat{b}_i$ to minimize it so that the outputs y_m and y_e would coincide as much as possible.

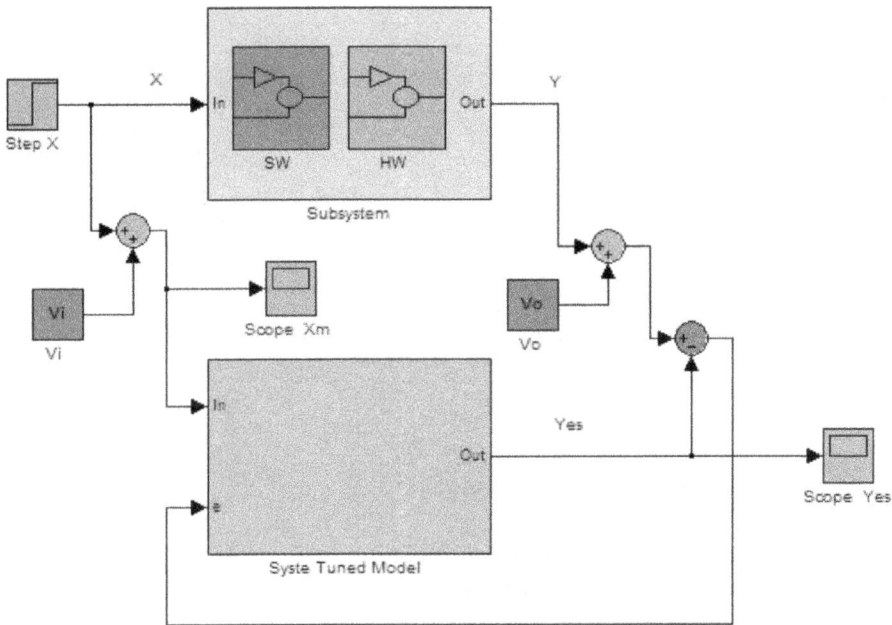

Figure 8.
System parameter identification experiment scheme.

It can be mentioned that such identification does not require simulating of system dynamic in closed feedback control loop configuration. To identify open-loop transfer function is enough, then closed-loop transfer function can be recalculated with the following formula [27]:

$$W(s) = \frac{G(s)}{1 + G(s)} \tag{18}$$

where $W(s)$ is negative feedback control closed-loop transfer function, $G(s)$ is transfer function of this loop in open state (assuming that feedback has unit transfer function $F(s) = 1$).

This is important for SS and specifically for SCS because it does not require unique complex equipment to simulate space flight and closed feedback control loop formed by the SCS in it.

Basic ideas of such a simulation for the identification of transfer function of open loop of SCS are presented in **Figure 6**.

The flight model of SS is installed on laboratory AIT table and electrically connected to the Laboratory Control-verification console.

SS expected transfer function $G(s)$ is known and should be verified with ITB, or in other words, its experimental estimate $\hat{G}(s)$ should be identified.

SS is switched on in special Ground Test Mode (GTM) (**Figure 9**). Its power, reference, and control data **D** are supplied *via* special data link from the laboratory Control and Verification Console (CVC).

It is important to note that in GTM SS should use special reference data about its state in SQ facility: Φ_0—latitude, Λ_0—longitude, h_0—altitude, $V_0 = 0$—velocity, B_0—magnetic induction vector. Its input is physically activated with a kind of laboratory imitator (red arrow in **Figure 6**). SS input and output data X_m and Y_m are recorded in real time in the CVC. After the end of the experiment, these data are

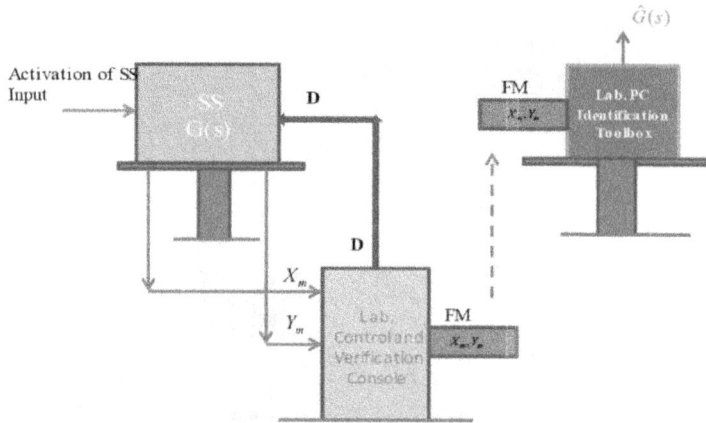

Figure 9.
Scheme of SS transfer function identification experiment.

reformatted in the form of mat. File and downloaded into the flash memory chip (FM in **Figure 6**) that using regular USB interface is connected to laboratory PC for the data post-processing in ITB. This ITB carries out the estimate of SS transfer function $\hat{G}(s)$. If it is close to expected due the SS design function $G(s)$, then we can allege that $G(s)$ is verified by SQ FT.

Examples of identification of basic dynamic units

With purpose to validate identification method for SS SQ FT before performing seminatural simulations, some typical liner time-invariant (LTI) dynamic unit transfer functions were identified with mathematical (quasi-seminatural simulation). Some examples can be also found in [28].

The same methodology for this "quasi seminatural simulation" was used. At first, system was simulated without measured errors, idealistic ("clear" measurements) input X and output Y and its step response Y was received. After input $X_m = u$ and output Y_m were distorted with superimposed Gaussian white noises, imitated measured errors and these signals were used for identification system dynamics (transfer function, step response, amplitude/phase frequency diagrams, characteristic polynomial roots).

Example 1: Simplest aperiodic system, first-order unit.

Given system is first-order dynamic unit that has transfer function.

$$G(s) = \frac{1}{Ts + 1} \qquad (19)$$

where $T = 10$ s is system time constant.

Characteristic equation $Ts + 1 = 0$ root is $s_* = -\frac{1}{T} = -0.1 \, \text{s}^{-1}$.

Simulink block diagram of this system is presented in **Figure 10**.

This scheme allows analyzing the step response of the system without and with measured noise.

1a-Mathematical simulation

Step response of the system (16) without noise is shown in **Figure 11**.

1b-"Quasi semi-natural" simulation

Step response of the system (19) with noise is shown in **Figure 12**.

1c- Identification results

System (19) identification results are presented in **Figures 13–15**.

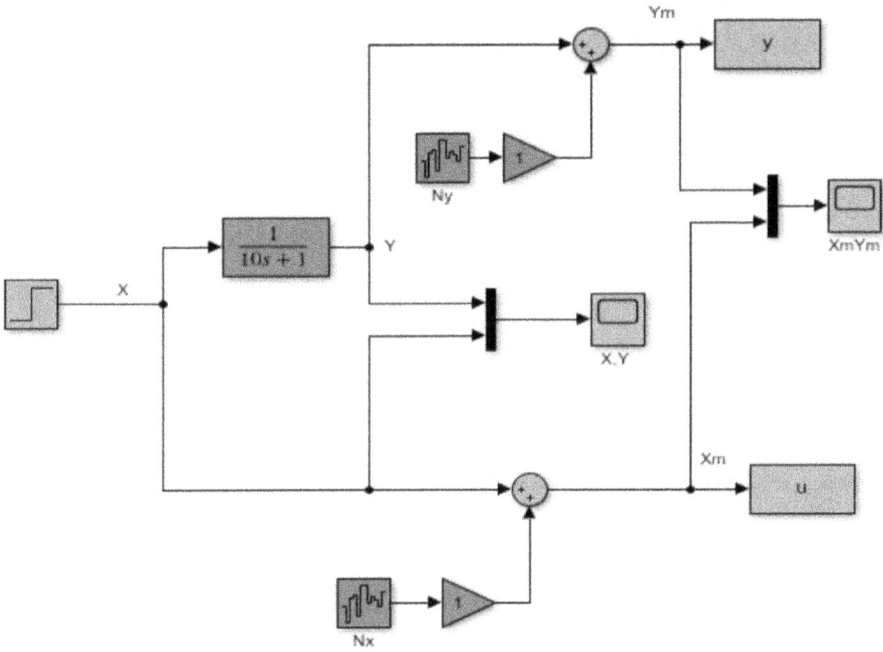

Figure 10.
Simulink block diagram of first-order unit.

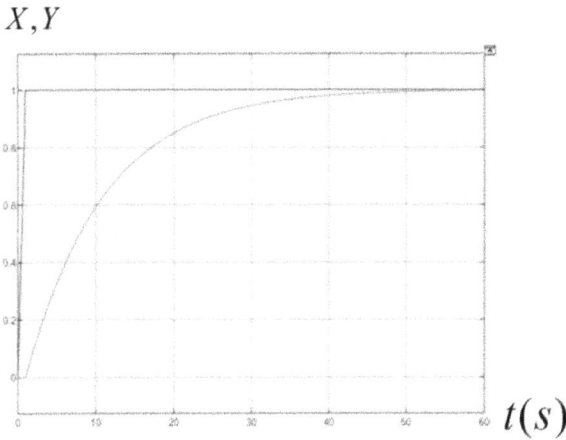

Figure 11.
Step response of the system (19) without noise. X is input-blue, Y is output-red.

Estimated characteristic equation of the system (12) is $\hat{T}s + 1 = 0$ with the root $s_* = -0.09$, estimated Time constant is $\hat{T} = \frac{1}{s_*} = 11.1$ s.

Estimated transfer function of the system (19) is

$$G(s) = \frac{1}{\hat{T}s + 1} \tag{20}$$

X_m, Y_m

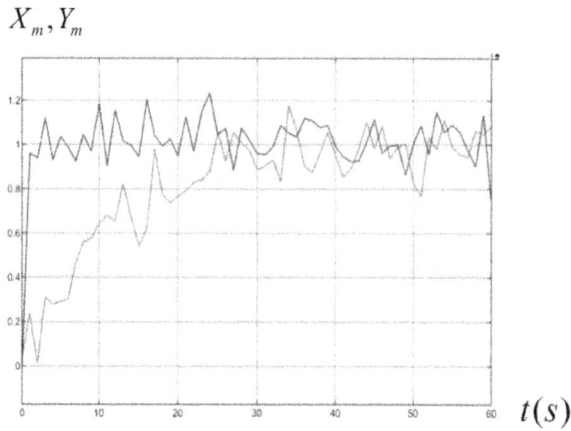

$t(s)$

Figure 12.
Step response of the system (19) with noise in measurements. X_m is input-blue, Y_m is output-red.

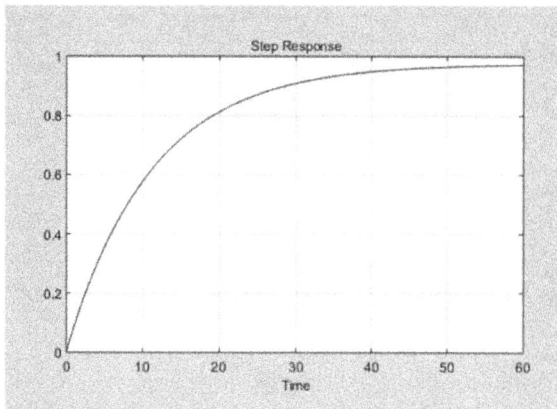

Figure 13.
Step response $h(t)$ of the identified system (19).

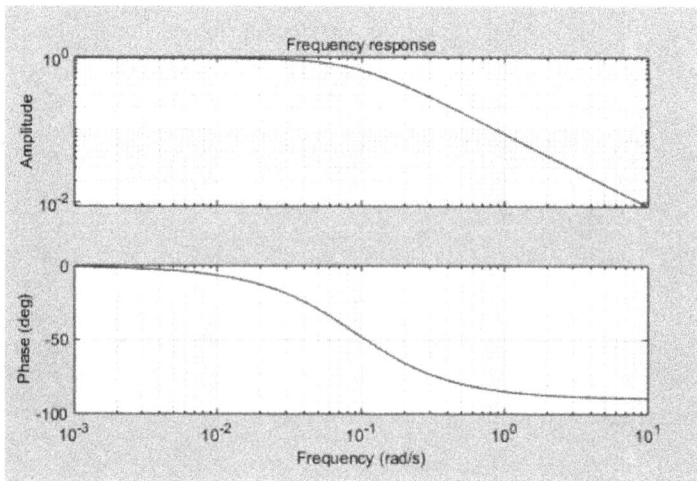

Figure 14.
Amplitude $A(\omega)$ and phase $\varphi(\omega)$ diagrams of the identified system (19).

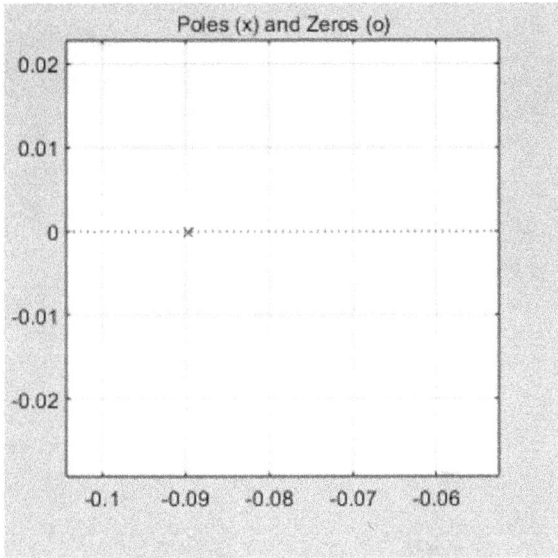

Figure 15.
Root of the characteristic equation of the identified system (19).

Example 2: Damped oscillator, second-order unit
Given system is second-order dynamic unit that has transfer function

$$G(s) = \frac{k}{T^2 s^2 + 2dTs + 1} \tag{21}$$

where $T = 10$ s is system time constant, $d = 0.707$ is specific damping coefficient, $k = 5$ is static control gain.

System characteristic equation is $T^2 s^2 + 2dTs + 1 = 0$. Its roots are $s_{*1,2} = -0.0707 \pm 0.0714j$.

Simulink block diagram of this system is presented in **Figure 16**.

This scheme allows analyzing the step response of the system without and with measured noise.

2a—Mathematical simulation

Step response of the system (21) without noise is shown in **Figure 17**.

2b—"Quasi semi-natural" simulation

Step response of the system (21) with noise is shown in **Figure 18**.

2c—Identification results

System (21) identification results are presented in **Figures 19–21**.

Estimated characteristic equation of the system (14) is $\hat{T}^2 s^2 + 2\hat{d}\hat{T}s + 1 = 0$. It has two complex roots $s_{*1,2} = -0.0622 \pm 0.0688i$,

Estimated transfer function of the system (21) is.

$$G(s) = \frac{\hat{k}}{\hat{T}^2 s^2 + 2\hat{d}\hat{T}s + 1} \tag{22}$$

where estimated Time constant is $\hat{T} = 10.78$ s, specific damping coefficient is $\hat{d} = 0.6707$, static control gain is $\hat{k} = 4.99$.

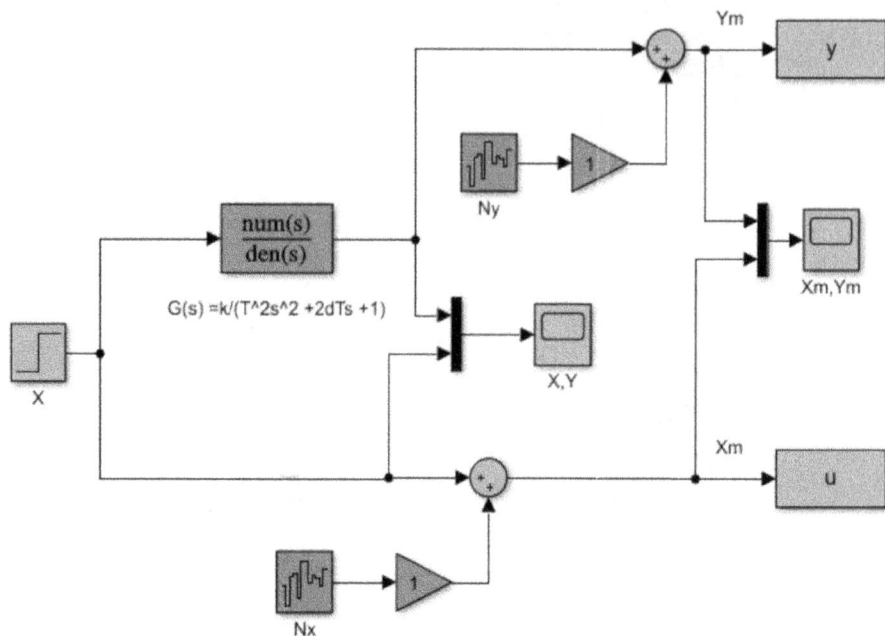

Figure 16.
Simulink block diagram of second-order unit.

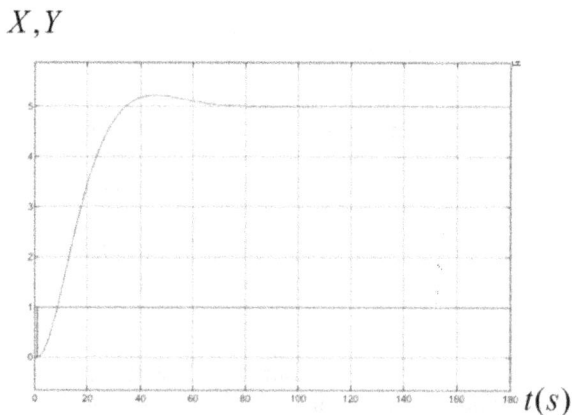

Figure 17.
Step response of the system (21) without noise. X is input-blue, Y is output-red.

Example 3: PID controller
    Given system is Proportional, Integral, and Damping controller that has transfer function

$$G(s) = k_p + \frac{k_i}{s} + k_d s \tag{23}$$

where the control gains are as follows: k_p is positional gain, k_i is integral gain, k_d is damping gain.

X_m, Y_m

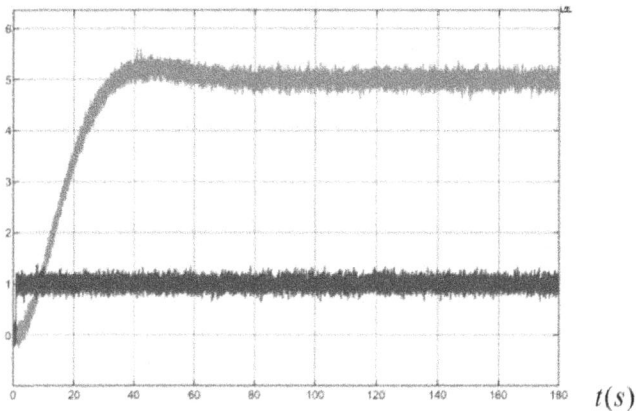

$t(s)$

Figure 18.
Step response of the system (18) with noise in measurements. X_m is input-blue, Y_m is output-red.

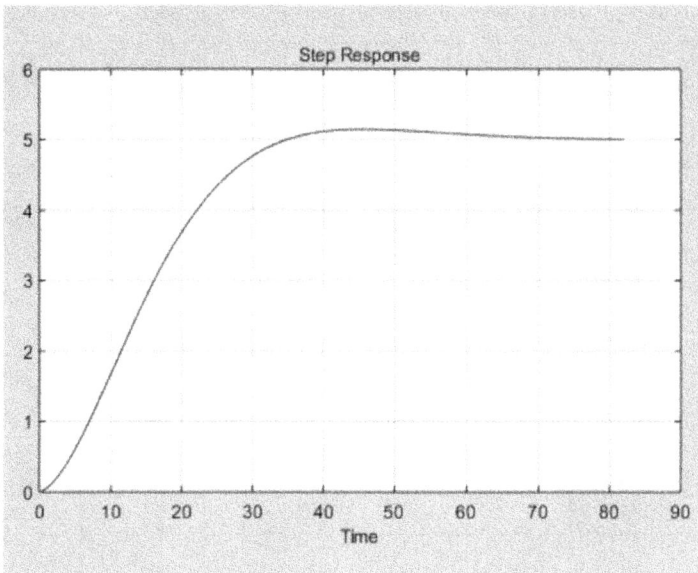

Figure 19.
Step response $h(t)$ of the identified system (21).

Practically, ideal differentiation assumed in (23) cannot be realized. Realistically, (23) should be represented as

$$G_c(s) = k_p + \frac{k_i}{s} + \frac{k_d s}{\tau s + 1} \tag{24}$$

where τ is a small time constant. In other words, the differentiation with filtering takes place and $\omega_c = \frac{1}{\tau}$ is the cut frequency (bandwidth) of this differentiating filter. Let us given, that
$k_p = 0.1Nm/rad, k_d = 0.03Nm/rad/s, k_i = 0.05Nm/rad \cdot s, \quad \tau = 10s$ (assuming that the input of this controller is an angle in radians—*rad*and output is the control torque in Newton meters—*Nm*).

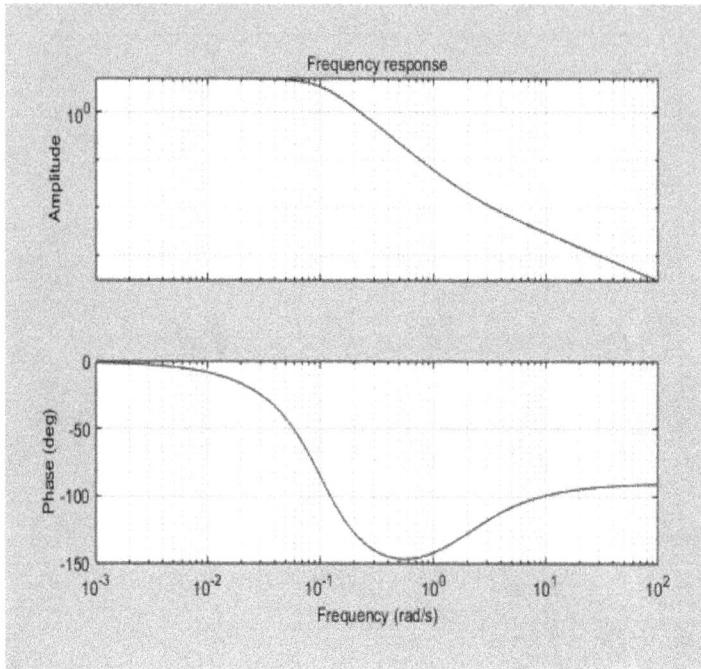

Figure 20.
Amplitude $A(\omega)$ and phase $\varphi(\omega)$ diagrams of the identified system (21).

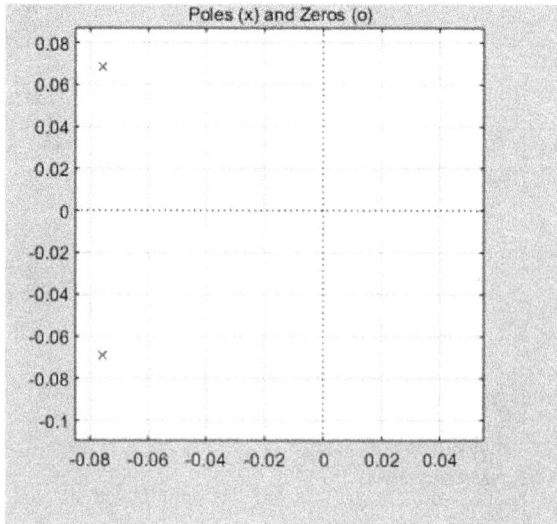

Figure 21.
Roots of the characteristic equation of the identified system (21).

After algebraic transformation (24) can be represented as follows

$$G(s) = \frac{\left(k_p \tau + k_d\right)s^2 + \left(k_p + k_i \tau\right)s + k_i}{s(\tau s + 1)} \tag{25}$$

or in the numerical form

$$G(s) = \frac{1.03s^2 + 0.6s + 0.05}{s(10s + 1)} \tag{26}$$

Denominator of (23) $s(10s + 1) = 0$ has following roots (poles): $s_{*1} = 0, s_{*2} = -0.1$ and the nominator $1.03s^2 + 0.6s + 0.05 = 0$ following (nulls) $s_1^* = -0.482$, $s_2^* = -0.101$.

Simulink block diagram of this PID controller is presented in **Figure 22**.
3a—Mathematical simulation
Step response of the system (24) without noise is shown in **Figure 23**.
3b—"Quasi semi-natural" simulation
Step response of the system (24) with noise is shown in **Figure 24**.
3c—Identification results
Identification results of the system (24) are presented in **Figures 25–27**.
Estimated transfer function of the system (23)/(24) is

$$G(s) = \frac{1.034s^2 + 0.0618s + 0.004929}{s^2 + 0.0986s + 1.289 \cdot 10^{-16}} \tag{27}$$

Formula (27) can be approximately represented as follows:

$$\hat{G}(s) \approx \frac{1.0487s^2 + 0.6103s + 0.05}{s(10.142s + 1)} \tag{28}$$

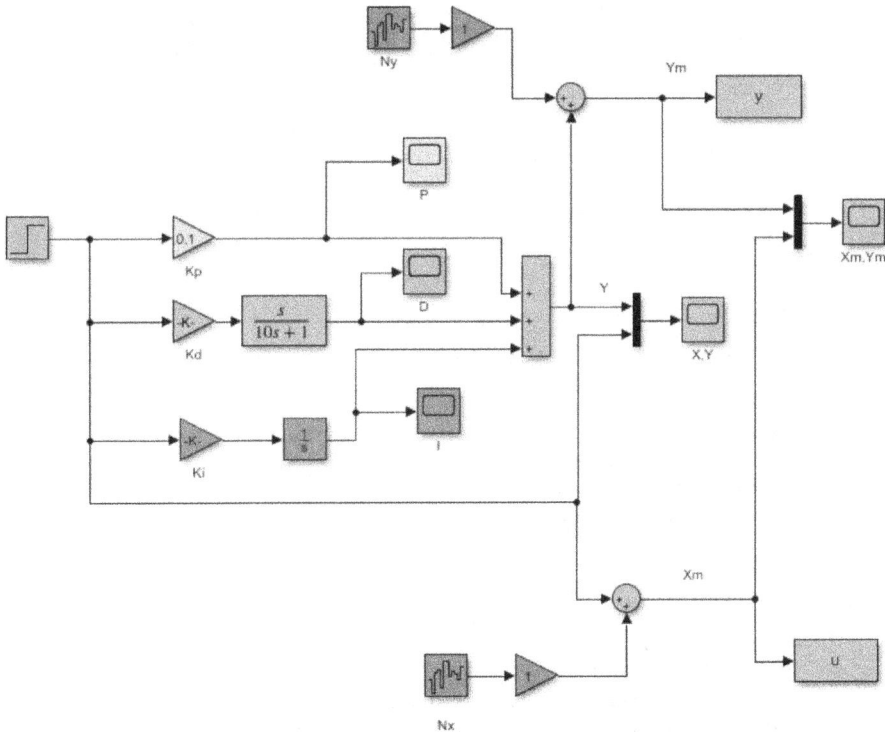

Figure 22.
Simulink block diagram of PID controller.

X, Y

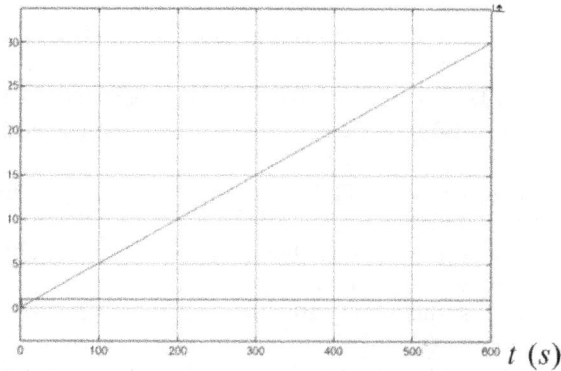

Figure 23.
Step response of the system (24) without noise. X is input-blue, Y is output-red.

X_m, Y_m

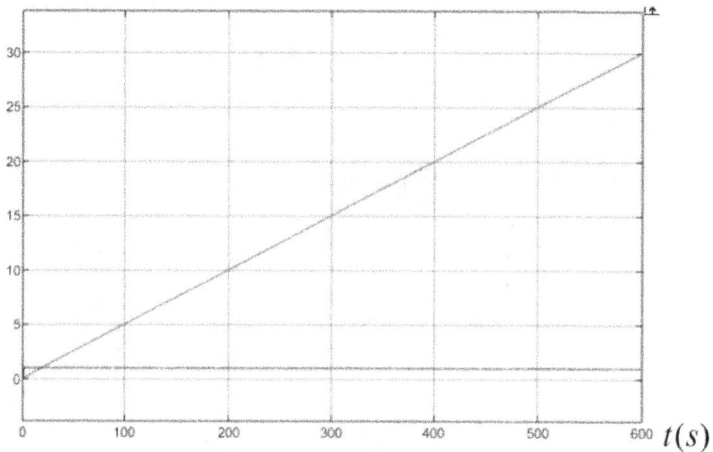

Figure 24.
Step response of the system (24) with noise. X_m is input-blue, Y_m is output-red.

Denominator of (28) $s(10.142s + 1) = 0$ has following roots (poles): $s_{*1} = 0, s_{*2} = -0.0986$ and the nominator $1.0487s^2 + 0.6103s + 0.05 = 0$ following (nulls) $s_1^* = -0.4818$, $s_2^* = -0.1008$.

Comparing coefficients (28) with (24), we can determine PID control gains and the time constant

$$k_p = 0.1033Nm/rad, k_d = 0.025Nm/rad/s, k_i = 0.05Nm/rad \cdot s, \quad \tau = 10.142s$$

Comparing identification results obtained with "quasi seminatural" simulation with real mathematical model, we can see that for all three considered above examples, identified model takes place in close coincidence between real and identified models that show effectiveness of application of ITB for identification purposes.

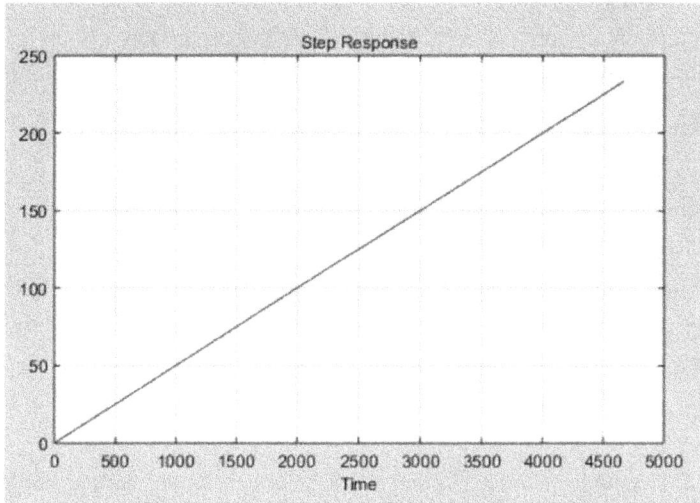

Figure 25.
Step response h(t) of the identified system (24).

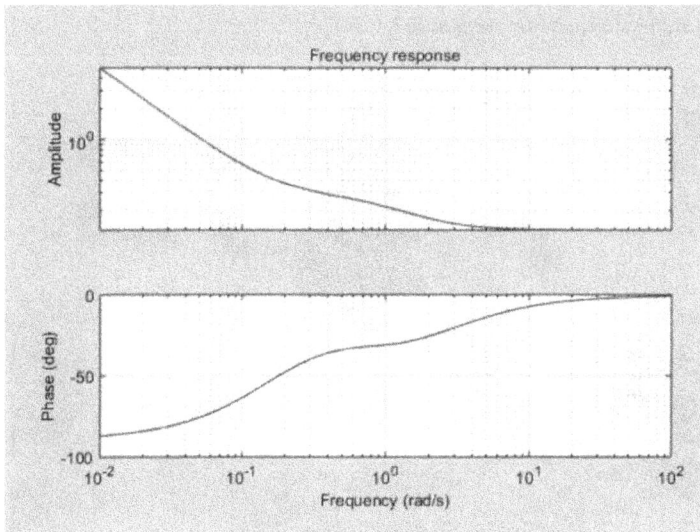

Figure 26.
Amplitude A(ω) and phase φ(ω) diagrams of the identified system (24).

Presented above examples show that Matlab Identification Toolbox, at least for simple LTI units, can be successfully used for identification their dynamic characteristics. Further studies should verify mathematical simulation with real physical experiments (semi-natural simulation), involving system hardware. More complex, nonlinear, and nonstationary systems also should be studied.

Related to these tests methodology, requirements and standards can be found in [8, 14, 18, 19, 29].

Practically, implementation of the presented above functional Space Qualification Test can essentially decrease of many unexpected flight anomalies that occurred and were learned during operation of first Canadian SSAR satellite Radarsat-1 (**Figure 28**).

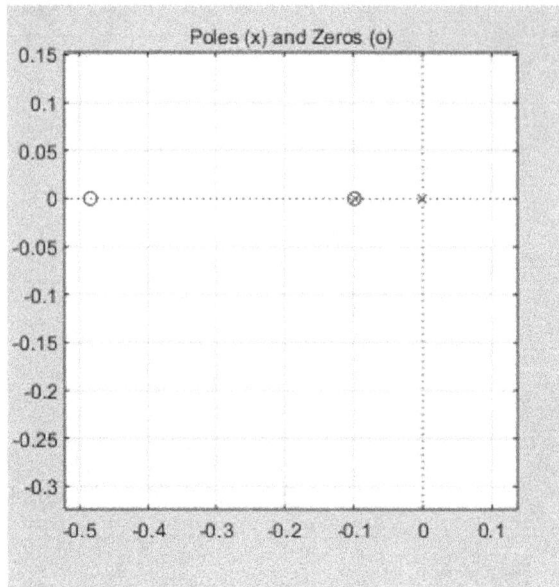

Figure 27.
Roots of the characteristic equation of the identified system (24).

Figure 28.
Canadian satellite Radarsat 2 in CSA David Florida space qualification test laboratory. Copyright: CSA//
www.asc-csa.gc.ca.

5. Conclusion

This chapter (Part II) continues (see Part I in [9]) to present a basic ground for Satellite Control System to integrate it with such a payload as satellite on-board SSAR RLS. Namely, it presents SCS: Control Modes, Power, Interface, and Testing. This material presented from the point of view of integration both systems into the satellite bus, considering satellite as the integration platform and seeing the satellite designer as the Prime Contractor, responsible for Earth observation mission successful execution in Space. However, in some cases the payload (for example, SAR) provider can perform the integration function as well. Of special interest can be, presented above, methodology of SCS Space Qualification (SQ) Functional Test (FT) that can be similarly applied to the remote sensing payload also and, finally, to the integrated system identifying its dynamic at the final stage of satellite Space Qualification program.

The chapter can serve to a wide pool of Space system specialists as an introduction to Satellite Control System development.

Acknowledgements

The author wishes to express his sincere gratitude to the Canadian Space Agency, where he had the opportunity to learn and possess the knowledge and experience related to the writing of this chapter. In addition, he is very thankful to many of his colleagues from Canadian Magellan Aerospace Company (Bristol Aerospace Division) with whom he discussed and analyzed satellite AODCS design projects and issues that helped him to work out the system analysis and its principal concepts presented in this chapter. Additionally, he cannot forget that his experience and background in Aerospace Technology, which were also accumulated from the former USSR (Moscow Aviation Institute, Moscow Aviapribor Corporation, Moscow Experimental Design Bureau Mars, Institute in Problems in Mechanic of RAN) and Israel (IAI, Lahav Division and Tashan Engineering Center), where he could observe and learn from diverse and wealthy engineering and scientific schools led by great scientists and designers such as Prof. B.A. Riabov, Prof. V.P. Seleznev, Prof. V.E. Melnikov, V.A. Yakovlev, G. I. Chesnokov, Dr.V.V. Smirnov, Dr. A. Syrov, Acad. F. Chernousko, A. Sadot, and Dr. I. Soroka.

This chapter was written as a solo author since his friend and regular coauthor Prof. George Vukovich from York University of Toronto passed away 4 years ago. For many years, Prof. Vukovich served as Director of the Department of Spacecraft Engineering in CSA. He will always keep good memories about Prof. Vukovich, who helped and encouraged him to continue his scientific and engineering work.

The author also acknowledges the copyrights of all publishers of the illustrations that were extracted from the open sources in the Internet.

Author details

Yuri V. Kim†
David Florida Laboratory, Canadian Space Agency (CSA), St. Hubert, Ottawa, Canada

*Address all correspondence to: yurikim@hotmail.ca

† Note: Dedicated to Prof. G. Vukovich.

IntechOpen

References

[1] Hudson B. Syntetic Aperture Radar. Concept of Operation, CC BY SA 4.0, Open Source Satelite, opensoursesatellite.org. Available from: https://www.opensourcesatellite.org. [Accessed: December 2021]

[2] Wachon P, Krogstad HE, Paterson J. Airborne and Space borne SAR Observation of Ocean Waves. Available from: https://www.researchgate.net/de ref/http%3A%2F%2Fwww.tandfonline. com%2Fpage%2Fterms-and-conditions. [Accessed: December 2021]

[3] Liang J, Zhang H, et al. Study on pointing accuracy, effect on image quality of space-borne video SAR. In: IOP Conference Series: Materials Science and Engineering. Vol. 490 (072011). IOP Publising; 2019

[4] Verba VS, Neronsky LB, Turuk VE. Spacebased Radiolocation Systems of Earth Observation. Radiotechnika, (rus.) M; 2010

[5] US National Strategy for Civil Earth Observation. Washington, DC: Executive Office of US President Science and Technology Council; 2013

[6] Rees WG. Physical Principles of Remote Sensing. UK: Cambridge University Press; 2001

[7] Sidi M. Spacecraft Dynamics and Control. A Practical Engineering Approach. Cambridge: Cambridge University Press; 1997

[8] Technical Committee ISO/TC 20. Space Systems, Design, Qualification and Acceptance Tests of Small Spacecraft and Units, International Standard, ISO 19683. Geneva, Switzerland: ISO (International Standard Organization); 2017

[9] Kim YV. Satellite control system: Part I – Architecture and main components, Section 2. In: Nguyen T, editor. Satellite Systems; Design, Modeling, Simulation and Analysis. London: IntechOpen; 2021. pp. 45-78 Available from: http://www.intechopen. com/books/satellite-systems-design-modeling-simulation-and-analysis

[10] Srivastava SK, Cote S, Le Dantec P. RADARSAT-1 image quality excellence in the extended mission satellite operations. In: Geosciences and Remote Sensing Symposium 2005, IGARSS '05. Vol. 1. IEEE International; 2005

[11] Bryson AE, Yu-Chi Ho J. Applied Optimal Control. Levittown, PA: Taylor & Francis; 1975. pp. 364-373

[12] Kim YV. Kalman filter and satellite attitude control system analytical design. International Journal Space Science and Engineering. 2020;**6**(1): 82-103

[13] MathWorks, Inc. Training course. Adopting Model Based Design. Natick, MA: MathWorks, Inc; 2005

[14] Space Engineering, Satellite Attitude and Orbital Control System (AOCS) Requirements, ECSS-E-ST-60-30C. 2013

[15] Eickhoff J. On Board Computers, On Board Software and Satellite Operation. Berlin: Springer-Verlag; 2012

[16] US Department of Defense. MIL-STD-1553B, Digital Time Division Command/Response Multiplex Data Bus. 1987

[17] ANSI (American National Standards Institute). RS-422 and RS-485 Application eBook (A Practical Guide). Washington, DC: International Headquarters Mfg. Co. Inc; 2010.

Available from: http://www.bb-elec.c
om/Learning-Center/All-White-Papers/
Serial/RS-422-and-RS-485-Applica
tions-eBook/RS422-RS485-Application-
Guide-Ebook.pdf. [Accessed:
February 13, 2020]

[18] Welch J. Flight Unit Qualification
Guidelines, Aerospace Report
TOR-2010(8591)-20. CA: US AF&
AEROSPCE Ltd. (Assuring Space
Mission Success); 2010

[19] ECSS Testing Working Group.
Space Engineering, Testing,
ECSS-E-10-03A, ESA-ESTEC.
Noordwijk: ECSS (European
Corporation for Space Standardization)
Publication Division; 2002

[20] Dennehy CJ et al. GN&C
Engineering best practices for human
rated spacecraft system. In: AIAA
Guidance, Navigation and Control
Conference and Exhibit. California:
AIAA2007-6336 (on line); 2007. DOI:
10.2514/6.2007-6336. [Accessed:
January 2021]

[21] Gavigan P. Design, test, calibration
and qualification of satellite sun sensors,
power systems and supporting software
development [thesis M.Sc.]. Toronto,
ON: University of Toronto; 2011

[22] Crowell C. Development and
analysis of a small satellite attitude
determination and control system
testbed [thesis M.Sc.]. MA: MIT; 2011

[23] Wertz J. Spacecraft Attitude
Determination and Control. Dorderiht:
Kluwer Academic Publisher; 1978

[24] Kim YV. Verification of space system
dynamics using the MATLAB
identification toolbox in space
qualification test. International Journal of
Aerospace and Mechanical Engineering
(Dubai). 2021;15(8):360-367

[25] Eykhoff P. System Identification,
Parameter and State Estimation. NJ, US:
J.Wiley and Sons Ltd.; 1974

[26] Ljung L. System Identification
Toolbox for Use with Matlab Guide.
Natick: MathWorks, Inc.; 2014.
Available from: www.researchgate.net/
publication/37405937. [Accessed:
December 2020]

[27] Aström KJ. Ch. 5. In: Control System
Design. CA: University Santa Barbara;
2002. pp. 181-182. Available from:
www.cds.caltech.edu/cds101/astrom.
[Accessed: January 2021]

[28] Fruk M, Vujisie G, Spoljarie T.
Parameter identification of transfer
functions using Matlab. In: Proceedings
of 2013 36-th International Convention
on Information and Communication.
Vol. 2(1). NJ, US: IEEE; 2013.
pp. 571-576. [Accessed: January 2021]

[29] Holodkov N, editor. Experimental
Refinement of Space Flight Apparatuses
(rus.) Moscow: MAI, 30; 1994

www.ingramcontent.com/pod-product-compliance
Lightning Source LLC
Chambersburg PA
CBHW081543190326
41458CB00015B/5629